The Climate Change Debate

Recent Titles in the

CONTEMPORARY WORLD ISSUES
Series

Books in the **Contemporary World Issues** series address vital issues in today's society such as genetic engineering, pollution, and biodiversity. Written by professional writers, scholars, and nonacademic experts, these books are authoritative, clearly written, up-to-date, and objective. They provide a good starting point for research by high school and college students, scholars, and general readers as well as by legislators, businesspeople, activists, and others.

Each book, carefully organized and easy to use, contains an overview of the subject, a detailed chronology, biographical sketches, facts and data and/or documents and other primary source material, a forum of authoritative perspective essays, annotated lists of print and nonprint resources, and an index.

Readers of books in the Contemporary World Issues series will find the information they need in order to have a better understanding of the social, political, environmental, and economic issues facing the world today.

The Climate Change Debate

A REFERENCE HANDBOOK

David E. Newton

ABC-CLIO®

An Imprint of ABC-CLIO, LLC
Santa Barbara, California • Denver, Colorado

Library of Congress Cataloging-in-Publication Data

Names: Newton, David E., author.
Title: The climate change debate : a reference handbook / David E. Newton.
Description: Santa Barbara, California : ABC-CLIO, an Imprint of ABC-CLIO, LLC, [2020] | Series: Contemporary world issues | Includes bibliographical references and index.
Identifiers: LCCN 2019051601 (print) | LCCN 2019051602 (ebook) | ISBN 9781440875410 (hardcover) | ISBN 9781440875427 (ebook)
Subjects: LCSH: Climatic changes.
Classification: LCC QC903 .N495 2020 (print) | LCC QC903 (ebook) | DDC 363.738/74—dc23
LC record available at https://lccn.loc.gov/2019051601
LC ebook record available at https://lccn.loc.gov/2019051602

ISBN: 978-1-4408-7541-0 (print)
     978-1-4408-7542-7 (ebook)

24 23 22 21 20   1 2 3 4 5

This book is also available as an eBook.

ABC-CLIO
An Imprint of ABC-CLIO, LLC

ABC-CLIO, LLC
147 Castilian Drive
Santa Barbara, California 93117
www.abc-clio.com

This book is printed on acid-free paper ∞

Manufactured in the United States of America

Contents

June 2081

"Hey, Mom. I have a question for you. Grandma's been telling us stories about 'the good old days,' you know, back in the 2020s. And I've been reading about and looking at pictures of those times. How come so many things seem so different today than they did in the 2020s?

"Like the winters, for example. Grandma showed us a picture of big snow piles around her house when she was a little girl. We still have snow, of course, but I've never seen so much all at once in our town. And speaking of snow, what happened to those huge fields of snow and ice that I see in pictures of the Alps and Andes and Himalayas? Sure, they still have snow and ice too, but it's nothing like we see in pictures today. And what about Greenland and the Arctic? Maps of those areas today look so different than they did in the 2020s. It looks as if someone has colored large parts of the white in those maps with a green crayon. Why trees and bushes now rather than ice and snow then?

"And other things too. Like that storm that wiped out San Juan last fall. Sure, hurricanes and tornadoes have been around forever. But how come they seem to be so much worse now than in Grandma's day?

"I love to look at the maps from the old times. Like the one from 2015 that shows Ocean City as being at the end of a long peninsula. Today, Ocean City is an island! What' with that?

"Even our own backyard looks different. Grandma told us about the vegetable gardens she used to grow to have fresh produce all summer. We tried to grow tomato plants last summer, and they all died within a couple of weeks. Even the trees and bushes around our house look different from the ones in Grandma's pictures. What happened to those trees and bushes, and where did the new ones come from? Same way with birds and other animals. Some of the ones in her pictures don't even exist around here anymore. Where did they go?

"I was especially curious about vaccinations for kids like me. Grandma talks about 'measles and mumps' a lot, but we don't even have those diseases any more. Instead, we have to have vaccinations for dengue fever and leishmaniasis, and Grandma says she's never heard of those diseases.

"And how about the price of food? Grandma said that she could buy an apple for only 30 cents when she was a girl. We could never get one for less than about 75 cents today. Even the economy seems to have changed from 2020 to 2080.

"What confuses me most is that nobody in the 2020s seemed to know anything about the possibility that these changes might occur. Or, if they did, they didn't do anything about it. Yet, my science teacher tells us that people knew the climate change was coming at least 200 years ago. Did people just ignore what scientists already knew, and that they were trying to tell the world? Or did government officials, leaders of industry, church leaders, and ordinary people just decide that the climate change story was nothing to worry about? Or that it would cost too much money to do something about it? Why didn't people act more aggressively while they still had that chance. Because it's too late for us to do much today."

"Well, Robin, that's a long and complicated story . . ."

And we'll try to tell the basics of that story in this book. Yes, scientists have had at least a hint of possible climate changes since at least the end of the nineteenth century. And more and more evidence of global warming really began to accumulate by the second half of the twentieth century. But, as of 2020, efforts to take actions to prevent, delay, and/or deal with climate change were still quite meager. A handful of nations, regions, and states have tried to create programs of one type or another to deal with climate change. But the problem is much too large for that approach. It's the world we're talking about here; not a handful of nations, regions, and states.

Chapter 1 of this book is designed to provide the reader with a general history of climate science. It describes some of the seminal discoveries about the relationship of solar energy, carbon dioxide and other greenhouse gases in the atmosphere, global temperature changes, and changes in the natural and human environment likely to occur as a result of these events. The chapter also attempts to provide a general introduction to the contrast between developments in climate science research and the response by governments, business, and the general public to the news coming out of that research.

Chapter 2 focuses on some of the specific consequences that might be expected from climate change and the way those consequences could affect the natural environment and human health. A fundamental element of that discussion is the research conducted and disseminated by the Intergovernmental Panel on Climate Change, formed in 1988. By 2020, the IPCC had issued five complete reports on climate change and had started on its sixth report, due in 2022. They are an invaluable resource in gaining a solid understanding of the climate change issue in the world today.

Chapter 2 also reviews the phenomenon known as climate skepticism or climate denial. These terms refer to the fact that, while climate change is now almost universally accepted as a reality by researchers in the field, a substantial number of politicians, government leaders, religious leaders, and everyday citizens continue to deny that climate change exists, that humans have any role in the phenomenon, that it will have any meaningful impact on human life, and/or that it will be impossible or economically undesirable to take action to prevent future climate change.

Remaining chapters in the book provide additional resources to assist the reader with a better understanding of the climate change issue and to help guide one's future research in the field. These chapters provide an annotated bibliography, a chronology of events, biographical sketches of important individuals and organizations in the field, and a glossary of essential terms used in the field of climate science. Chapter 3 of the book, Perspectives, also provides a group of essays by interested individuals on topics of specific interest to those individuals.

The Climate Change Debate

"Global warming is the most pressing problem the world faces." (Benn 2007)

". . . we are as confident that humans cause climate change than that smoking causes cancer." (Ocko 2019)

"The climate has changed before. Humans were not around. Therefore, humans do not cause climate change." (Sense Seeker 2012)

"The concept of global warming was created by and for the Chinese in order to make U.S. manufacturing non-competitive." (Trump 2012)

Countless numbers of comments such as these can be found in the print, electronic media, and social media today and the recent past. Is the climate really changing? If so, do humans play any part in the changes that are taking place? What are some of the issues that a changing climate might create for Americans and the rest of the world? What solutions may be available for dealing with those issues? The subject of climate change involves so many questions that it is worth writing a book about! So, here it goes.

Two polar scientists work on ice cores over an ice floe in Antarctica. (Tenedos/Dreamstime.com)

The Climatic Record

The first question one might ask in this "climate change debate" is what we know about the topic from Earth's history. Is the changing climate observed today a new phenomenon? Or is it an integral part of Earth's history? If the latter, how do scientists explain the changes that have taken place in climate? How do they collect information about climatic patterns a hundred years ago; a thousand years ago; a million years ago (if at all)?

Weather and Climate

The topic of climate change is intimately related to the subject of global warming. Although climate is the product of many factors, Earth's annual average temperature is certainly one of the most important. If the planet's temperature increases by one degree Celsius over a 100-year period, that change will have an observable effect on seasons, weather patterns, sea level, biological diversity, and a host of other factors. So, in the simplest possible terms:

climate change = global temperature + other stuff

A brief aside: Temperature measurements appear frequently in this book. Just for the record:

A change of 1°C = a change of 1.8°F
(one degree Celsius)   (one degree Fahrenheit)

A change of 1°F = a change of 0.56°C

The first issue in this review requires a discussion of the distinction between *weather* and *climate*. One finds in many debates over climate change comments such as, "Well, this was the coldest winter in a hundred years in Smithville. So how can global warming be taking place?" "The coldest winter in a hundred years" refers to atmospheric conditions that take place in a single year: the *weather* for that year. Weather measurements usually cover a period of a few weeks, a few months, or a few years. Scientists know a great deal as to the factors that cause weather conditions and can predict with at least some degree of accuracy what the weather will be like in the short term.

By definition, the term *climate* refers to longer stretches of time, usually 30 years or more. Climate measurements differ from weather measurements because the former take into consideration fluctuations from year to year in weather patterns. So, yes, it might have been very cold in Smithville last winter, but it's possible the next five winters were somewhat warmer, about the same as usual for Smithville (What's the Difference between Weather and Climate? 2018).

Climate through the Millennia

Articles about climate change often include the phrase "since 1850" or "since 1880" or something similar. The reason for these qualifiers is that the data needed to develop records for weather patterns and, therefore, climatic patterns, did not exist to any great extent before the mid-19th century (Cable 2009; "History of the National Weather Service," n.d.). Studies of climate prior to this period have few or no human records on which to rely. They depend, instead, on so-called proxy records, that is, information obtained from other than human sources, such as ice cores, tree rings, boreholes, lake and ocean sediments, isotopic composition of water, and evidence of life forms such as corals, dinoflagellates, and microbial mats (for an understandable description of one type of proxy, see Evans 2013; for a detailed technical explanation and database of proxies, see "Chemical Proxies Finder" 2016; for a general overview of the topic, see Dewey 2010).

Here is just one example of a climate change proxy. O^{18}/O^{16} (oxygen-18/oxygen-16) analysis is a form of climate change proxy using two isotopes of the element oxygen. Isotopes are forms of an element that differ in their atomic weights. The atomic weight of O^{18} is 18 (as you can tell from the superscript), and that of O^{16} is 16. That means that an atom of O^{18} weighs a bit more than an atom of O^{16}.

It is fairly easy to measure the amount of "heavier" water (H_2O^{18}) and "lighter" water (H_2O^{16}) in any sample of ocean or lake water or ice taken from a glacier or ice cap at any time in history. Suppose someone made that measurement of ice in Greenland today and found the O^{18}/O^{16} ratio to be 0.00018. If someone in 2100 made that same measurement he or she might find a different O^{18}/O^{16} ratio, say 0.00023. The reason for that difference is that H_2O^{18} is slightly heavier than H_2O^{16}, so it will evaporate more slowly than H_2O^{16}. That difference tells us that the average global temperature has changed from 2020 to 2100: ice in glaciers and ice caps has lost H_2O^{16} faster than it has lost H_2O^{18}. The evaporated H_2O^{16} has formed clouds, from which rain falls over dry land. So, the composition of ice and liquid water between 2020 and 2100 provides information as to temperature changes that have occurred during that period.

Climate change proxies are an essential tool for paleoclimatologists. Paleoclimatologists are researchers who study the climate in the distant past. "Distant past" can mean anything from a few thousand to a few billion years ago. Paleoclimatology differs from modern meteorology in some significant ways. Today's meteorologist studies, reports on, and attempts to predict a host of factors that affect the weather: the temperature now and in the near future at a variety of locations; atmospheric pressures in those places; the size and movement of air masses and the fronts they produce; patterns of precipitation; the likelihood, size, and location of severe storms; air movements in the upper atmosphere; and so on. Paleoclimatologists have a much more limited set of data to draw from, essentially annual temperature patterns over a long stretch of time (usually many decades or centuries), and water and ice patterns across large parts of the globe. The charts and graphs that appear in paleoclimatological reports, then, usually provide data about large-scale temperature and moisture patterns. And that's about it.

One way in which paleoclimatologists use the O^{18}/O^{16} proxy is as follows. First, they bore down into an ice field or glacier 100 feet, 200 feet, 500 feet, or greater depths and collect core samples of ice from each depth. The O^{18}/O^{16} ratio at each depth tells what the relative temperatures were at each

historical period during which the ice formed. This information allows the paleoclimatologist to construct graphs (for an
example, see Watts 2013a, Geological Timescale: Concentration of CO_2 and Temperature Fluctuations; for a similar application of the O^{18}/O^{16}, see Pearson 2012).

It is probably obvious that the further one goes back in history, the less confidence one can have about the data and other
information collected from proxy sources and the less certain
one can be about Earth's climate. In the first few billion years
of existence, therefore, relatively little is known with any certainty about Earth's atmosphere. In fact, no atmosphere of any
kind (and, therefore, any climate of any kind) would have been
possible The Earth was so hot and so early in its development
that whatever atmosphere there was consisted of the gases from
which the planet was made, hydrogen and helium, which rapidly escaped into outer space.

About four billion years ago, the first elements of a climate
system began to form. Water released from Earth's surface
began to escape into the atmosphere, eventually forming water
clouds. Those clouds evolved until they were large enough to
drop water on Earth's surface, forming the first oceans. The existence of oceans then made possible the development of the
earliest life forms (Hessler 2011; McLamb 2011).

Beginning about 2.4 billion years ago (bya) (also 2.4 gya for
"giga") ice began to form across Earth's surface, producing the
first true "ice age" in Earth's history. Since that time, overall
a total of five ice ages have been thought to exist. Those five
ice ages are as follows. (There is considerable variation in the
naming of ice age periods in various parts of the world and by
different researchers, as well as differences of opinion about the
timing of the ice age eras.)

Huronian Glaciation

This period of glaciation is thought to have lasted until about
2.1 bya. Unicellular organisms that had first appeared prior to
the Huronian period seem to have been completely destroyed

by the most extensive and longest-lasting glacial period in Earth's history. Some evidence suggests that volcanic activity sharply decreased during this period, reducing the amount of carbon dioxide released to the atmosphere and, hence, the atmosphere's ability to retain heat (Bekker 2014).

The Huronian ice age was followed by a very long period of time during which temperatures increased and glaciers and ice caps decreased in size. Periods of this type are called *interglacial periods*, or just *interglacials*. The five ice ages discussed here were all separated in time by interglacial periods. Different methods for naming interglacials are used, the most common of which is called the Marine Isotope Stages (MIS) method. That system is based on the levels of the O^{18}/O^{16} proxy in Foraminifera, marine microorganisms. According to that system, there have been 21 distinct interglacial periods, each of which may consist of two or more sub-periods (Berger et al. 2016; Railsback et al. 2015; for the relevant graph, see http://railsback.org /Fundamentals/SFMGSubstages01.pdf).

Snowball Earth

The second ice age is generally known as the Cryogenian period, or, more informally, Snowball Earth. It began about 720 million years ago (mya) and probably lasted until about 635 mya. During this period, the planet's average annual temperature may have dropped by as much as 12 degrees Celsius (22 degrees Fahrenheit). That change was sufficient to cause all liquid water on land to freeze, covering it with a layer of snow and ice. It may also have turned ocean water into slush (Kopp et al. 2005; Schriber 2015).

Andean-Saharan

The third ice age is thought to have begun about 460 mya and to have lasted for about 40 million years. It gets its name from the fact that low temperatures and glaciation appeared to have occurred primarily in the equatorial area of the planet, from the modern-day Saharan desert to the Andes Mountains of western

South America (the Andean-Saharan Ice Age, Ep 3 History of Climate Change CO$_2$ and Human Evolution 2018).

Late Paleozoic

The fourth ice age began about 360 mya and lasted for about 100 million years. It has a number of fascinating features, characterized by the development of glaciers and other icy areas across a landscape in which the first major landforms were beginning to develop and move across Earth's surface. The period is also thought to be the precursor of the fifth ice age, in which glaciation gradually disappeared and was replaced by the type of greenhouse environment characteristic of our planet today (Montanez and Poulsen 2013).

Quaternary

The fifth ice age dates from about 2.6 mya ago to the present day. A graph of temperature changes during the period should look very familiar; it consists of regular ups and downs of temperature over the whole period (Verheggen 2016; based on Snyder 2016). One period of time of special interest toward the end of this era is the so-called Little Ice Age. One reason the period is of interest is that it is so close to our own day. The other is that so much detailed information, in contrast to much older periods, is available about the age.

The Little Ice Age is thought to have begun in about 1300 and lasted for about 550 years. To the extent that the latter date is correct, one might conclude that the world is currently in a warming trend from the last ice age. The Little Ice Age followed a period of relative warm temperatures usually referred to as the Medieval Warm Period. It is characterized by the same pattern of "warming, then cooling, then warming, then cooling" trend characteristic of all ice age patterns discussed earlier (see the graphs at Climate 2017; Cook 2019).

Limited evidence suggests that a cooling climate occurred in almost all parts of the world during the Little Ice Age. But the most complete picture we have of the period comes from Europe, the North Atlantic, and North America. Decreasing temperatures had some devastating effects on many parts of society, essentially wiping out agricultural activities in many regions. Reduced food sources, in turn, led to famine and a reduction in population size in many regions of the continent. Mountain-based glaciers also moved to lower levels, creating long-lasting ice fields in areas where they traditionally did not occur. Many rivers froze over, turning them into "ice skating rinks" that have been memorialized in famous paintings of the time. Rainfall, snowstorms, and other severe weather also appear to have become more common in many parts of the world (Blom 2019; Fagan 2019).

The Tools of Paleoclimatology

Articles about climate science and climate change often contain references to conditions or events that occurred a few thousand years ago, a few million years ago, or even a few billion years ago. How can scientists possibly know what our planet was like that far back in history? The earliest written records of which we know date to about 3400 BCE. However, a variety of techniques are available for dating objects a few thousands of years prior to that time, including carbon-14 or potassium-40 dating, engraved rocks, and certain types of fossils (Marks 2019). For more ancient (a few hundreds of thousands of years or longer), other techniques have been developed for dating materials. These techniques include boreholes, ice cores, insect remains, pollen remains, and dendrochronology (study of tree rings).

Carbon Dating

Carbon (or *radiocarbon*) dating is based on the fact that two isotopes of the element carbon (C) exist (i.e., carbon-12 and carbon-14). Carbon-12 (^{12}C) is a stable isotope, meaning that it exists without change essentially forever. By contrast, carbon-14 (^{14}C) is unstable. It is radioactive, emitting a beta particle (β; electron), breaking down into an isotope of nitrogen, nitrogen-14 (^{12}N). The half-life of carbon-14 is about 5,700 years. The half-life of an isotope is the time it takes for half a sample of the isotope to decay (change into a new form). Hence, after a period of about 5,700 years, a sample of 10 grams of carbon-14 would decay into nitrogen-14, and only five grams of carbon-14 would remain. After an additional 5,700 years, only 2.5 grams of carbon-14 would be left and so on.

When any organic material (any material that contains the element carbon) is alive, it exchanges carbon isotopes with the atmosphere, keeping the ratio between carbon-12 and carbon-14 constant. After an organism dies or a material is buried, that exchange can no longer occur. The ratio between carbon-12 and carbon-14 begins to change as some carbon-14 changes into nitrogen. The change in ratios might, as an example, change from 100 to 1 at some time in the past. But, as time passes and the material is no longer in contact with the atmosphere, the ratio may change from 200 to 1, 400 to 1, 800 to 1, and so on. At some time, most of the carbon-14 has decayed to nitrogen-14, and it is no longer possible to measure the ratio between carbon-12 and carbon-14. This happens after about 50,000–60,000 years, the limit at which carbon dating can be used to find the age of an object or material (How Does Carbon Dating Work? 2019).

Other forms of radioisotope dating exist. For example, potassium-40 (^{40}K) is a radioactive isotope of the element potassium. It decays with the formation of the element argon, a gas. Rocks formed a few million or even a few billion years ago would have contained all isotopes of potassium, but no argon, which would have escaped from hot rocks as a gas to the atmosphere. As soon as the rocks cooled, however, potassium would be trapped within them and the decay of potassium-40 would have begun producing argon. The amount of argon found in a rock today, then, would be an indicator as to how long ago the rock would have been formed. Since potassium-40 has a half-life of about 1.248 billion years, the potassium/argon system can be used to date very early ages of Earth's history (Nave 2017).

Ice Cores

When geologists want to study Earth's early history, one technique they can use is to drill into rocks to various depths. The segments, "cores," collected from these drills show what Earth's crust looked like over hundreds or thousands of years. They may show times at which volcanoes were active, Earth's surface was moving, or other changes were occurring. Paleoclimatologists use a similar technique, except that they drill cores into ice sheets to recover cylinders of ice that provide evidence about climates at previous times in the past.

Most ice cores come from Antarctica or Greenland. In these regions, snow has been falling almost without interruption for thousands of years. Over time, the snow compacts to form ice and, in so doing, traps clues about Earth's climate at the time these events were occurring. For example, an ice crop may contain traces of a volcanic eruption or bubbles of gas. The bubbles can provide information about gases present in the atmosphere at the time the ice was forming. Some of the information provided by ice cores includes the rate at which snow and ice were falling at a given time; periods in which melting was occurring; other evidence of temperatures for each layer of an ice core; the amount of carbon dioxide, oxygen, and other gases present in the atmosphere during a time period; and the salinity of bodies of water within a core segment (About Ice Cores, n.d.).

Boreholes

Borehole technology is somewhat similar to ice core technology, with the exception that holes are drilled into the earth rather than glaciers or ice fields. The technology is based on the premise that heat of Earth's surface in the past is buried by accumulating layers of ground over time and that those surface layers retain their heat in the process. By drilling a hole in the ground to a depth of a few thousand feet, then, temperature measurements dating back up to about 1,000 years are possible.

Boreholes are accompanied by a number of technical problems that make them useful for temperature predictions only to a few hundred years in the past, sometimes approaching about 1,000 years ago. One problem is the continual warming of sub-surface regions by heat released from the planet's interior. Also, other factors may cause warming of regions adjacent to the borehole that cannot be traced to climates of the time. Researchers have developed sophisticated mathematical procedures, however, for adjusting with most of these problems. As a result, reasonably reliable data are now available from about 600 boreholes drilled in every part of Earth's surface, the greatest number of which have been drilled in Europe and North America. The drill holes have been extended to depths ranging anywhere from a few hundred to more than 3,000 feet into the earth (Huang, Pollack, and Shen 2000; Lemmons 2019).

Dendrochronology

Dendrochronology is the process of dating an object or material by examining tree rings. The technique was first developed in the early 20th century by American astronomer Andrew Ellicott Douglass. Douglass discovered that the characteristic features of tree rings corresponds to sunspot cycles. His research was later extended to archaeological objects and materials by American anthropologist Clark David Wissler. Wissler and Douglass developed a technique for using tree rings to date the object of a material to a point at least 10,000 years in the past.

Today, the science of dendrochronology makes use of a variety of detailed and sophisticated technologies to find the date of wood objects within a few decades for thousands of years in the past. The technology is based on the fact that all trees exhibit a common growth pattern: They grow more rapidly in the spring and summer, and more slowly in the fall and winter. These differences in growth rate are reflected by the size and color of trees ring. During growth spurts, rings tend to be significantly wider and light in color, and during periods of slow growth, narrower and darker. The problem for a dendrochronologist is that this general principle has many exceptions. For example, a tree near a river or lake typically receives all the water it needs year around, and rings tend to be more homogeneous in size and color. Trees in dry areas reflect this pattern, except rings are narrower and darker. In either case, the study of tree rings tells very little about weather patterns in the past. This analysis is also made more difficult because different tree species grow at different rates, resulting in different ring patterns. Thus, the rings in a 100-year-old oak tree tend to be different enough from those of a 100-year-old elm tree that comparing the two is difficult.

The standard method for using tree rings to determine the age of an object or material is called *cross dating*. The first step in cross dating is to study tree rings from a living plant. It is not necessary to cut the tree down to collect a ring specimen. Instead, one uses a boring tool to cut into a tree and remove a core sample. That sample is sufficient to examine the rings of the living tree (for a demonstration of the technology, see Novus RCN 2014). One can then simply count the rings in the tree to determine its age.

The second step in cross dating is to repeat this process with a dead tree. The dead tree is presumably older than the living tree, so the rings obtained date farther back. At this point, the two sets of tree rings can be overlaid. A match can then be made between certain segments of the two trees, allowing dating of the dead tree further back in time (for a diagram of this process, see Martinez [1996] 2000). This step can then be repeated at least once more, each time using a piece of wood older than the living tree and the dead tree. Repetition of this procedure can then be used to date an object back thousands of years, with information as to the weather and climatic conditions at the time (Sheppard 2014; Tree-Ring Dating 2013).

Pollen Analysis

Pollen is the fine, dusty material found in all flowering plants. It consists of the male gametes (sperm cells) used by plants to reproduce. Pollen particles are typically very small in size, ranging from about 10 to 200 micrometers (3×10^{-4} to 8×10^{-3} inches) in diameter. Researchers must use high-power microscopes to examine these particles. The outer walls of a pollen grain are made of a very tough material that is resistant to damage and decay in the environment. When pollen grains are blown into a lake, river, or other body of water, they usually sink to the bottom, are covered with silt, and eventually converted to sedimentary rock.

To determine the climate of an area using pollen, a paleoclimatologist must first recover pollen grains from some preserved environment, such as a sedimentary rock. The grains must then be removed from the source by physical and/or chemical means. The grains are then examined under a microscope, identified, and compared to pollen grains in the modern world or about which information is otherwise available. If a particular fossil pollen is found to match a pollen in existence today that comes only from plants that grow in very warm environments, one can conclude that the conditions under which the recovered pollen existed must also have been very warm. Since pollen-producing plants have existed on Earth for more than 300 million years, this climate proxy can be used to describe climate conditions very long periods into the past (Mason 2019; for an example of this line of research, see Zhao et al. 2014).

Pollen analysis is one division of a field of study called *palynology*, literally, "the study of dust." Besides pollen grains, palynologists study a variety of plant-based materials such as spores, orbicules, dinocysts, acritarchs, chitinozoans, and scolecodonts, together with particulate organic matter (POM) and kerogen. The methods used for these materials are similar to those used for pollen (Hirst 2019; Palynology n.d.).

Characteristic features of various unicellular microorganisms can also be used as climate proxies. The case of Foraminifera, discussed above, is an example. Foraminifera are protists that make a form of internal shell known as a *test* out of calcium carbonate, $CaCO_3$. The oxygen they use for this process consists of two isotopes of oxygen, ^{18}O and ^{16}O. As noted above, the ratio of these two isotopes, commonly known as $\delta\,^{18}O$, is an indicator of the time period during which the organism was making its tests and, hence, the age of the material in which it was found.

The physical structure of Foraminifera and other unicellular organisms also provides clues to the age in which they existed. The climate proxies that use dinoflagellates and diatoms are an example. The structure of these organisms found in some strata of rock can be compared to comparable members of a species alive today, and the climatic conditions of their origin can then be determined (Bhatia 2013; Pospelova, Pedersen, and Vernal 2006; Wetmore, n.d.).

A variety of other climate proxies are available for determining temperature and other atmospheric properties at distant times in the future. For a good general introduction, see Gornitz 2015; Paleoclimatology Datasets n.d. For a more detailed and more technical discussion of several climate proxies, see Climate Research Committee, National Research Council Staff 1996, ch. 5, 489–598.

Causes of Climate Change

Earth's climate is, and always has been, changing. That reality is made clear by the study of any graph of long-term temperature changes, from the planet's origin to the present day. Those graphs always consist of fluctuations from maxima to minima and back again (see Climate 2017) and never straight lines indicating no change in temperature. The obvious question is "why"? What are the factors that cause the average annual temperature on the planet to vary over time? The answer to that question is complex and involves many factors, some more important than others.

Milanković Theory

From a historical standpoint, the most important answer to that question comes from the research of Serbian geophysicist and astronomer Milutin Milanković (anglicized: Milankovitch). In the early 1920s, Milanković began an attempt to express changes in insolation (the amount of solar energy received by Earth over a period of time) in mathematical terms. He had already concluded that three fundamental factors were responsible for changes in insolation over tens and hundreds of thousands of years. Those factors were the *eccentricity* of Earth's orbit around the Sun; Earth's *obliquity*, its axial tilt; and *axial precession*, changes in the direction in which Earth's poles face against the starry background.

Eccentricity

Earth travels round the Sun in a nearly circular orbit. Its actual orbit is an ellipse, a flattened-out circle. The amount by which it is flattened is called its eccentricity. The eccentricity of Earth's orbit today is about 0.0167. By comparison, the eccentricity of a perfect circle is 0.0000. Earth's eccentricity varies from very nearly that of a perfect circle, 0.000055, to an ellipse with an eccentricity of 0.0679 over a period of about 413,000 years (Kershaw 2017; this text is an excellent reference for the review of astronomical factors affecting Earth's climate). Eccentricity influences Earth's climate because it affects the distance between Sun and Earth during different seasons. If the eccentricity is close to zero, all parts of the planet will be about the same distance from the Sun at all times of the year. As the eccentricity increases, Earth will be relatively close to the Sun at some times of the year, and further away at other times of the year. This difference is responsible for the seasons we experience on Earth. And the greater the eccentricity, the larger the difference in seasons, and the more likely the extreme temperatures needed for glaciation occur. This factor is, however, considerably less important than others in determining glacial conditions ("Is an Ice Age Coming?" 2016).

Obliquity

A more important influence on Earth's climate is the planet's obliquity, the amount by which Earth's axis is displaced from perpendicular to the direction of orbital motion. Earth's current obliquity is 23.44°. It varies over a period of about 41,000 years from 22.1° to 24.5° and back again. As obliquity decreases, winters tend to be less cold than at high obliquity and summers less warm. Warmer winter temperatures increase the rate at which water evaporates from Earth's surface, and hence, the amount of rain and snow that falls. In upper latitudes, snow and ice accumulate. At the same time, summer temperatures are less warm, so snow and ice formed during the winter is less likely to melt and evaporate. Conditions for glacial buildup improve, therefore, at low obliquities.

Axial Precession

The third factor involved in glaciation is axial precession, also known as precession of the equinoxes. If one could view Earth from outside our galaxy, the planet would look like a spinning top, with its axis pointed at some distant spot in the skies. If one had enough time—about 25,772 years—she or he would also notice that the axis would slowly trace out a circle in the skies to which it pointed. This change in the orientation of Earth's axis in space is called axial precession. As with eccentricity and obliquity, axial precession causes some parts of the planet to receive more or less insolation over long periods of time. Those with lower insolation and, therefore, lower temperatures, are more likely to be fruitful regions for the development of glaciation ("Milankovitch Cycles III: Precession of the Seasons, Seasonality, and Extent of Glaciation" 2006).

This explanation of the Milanković theory is a very simplified version of a very complex topic. Milanković's achievement was his ability to find a way of expressing these three factors in terms of mathematical equations. He found that times at which the three factors also acted in concert with each other, the likelihood of glaciation was significantly increased. (For a comparison of the combined effects and glacial conditions, see "Causes of Climate Change" [2019]; readers interested in a technical explanation of the Milanković theory should see Berger 1984.)

Other Possible Causes

Forces other than those discussed thus far are also responsible for the development of glacial periods. One of the most extensively studied of these forces is the movement of tectonic plates. Tectonic plates are massive blocks of rock that underlie Earth's crust. These plates are constantly in motion, causing them to collide with each other in a variety of ways. Researchers have identified or hypothesized a variety of ways in which tectonic movement may have contributed to the rise of the ice ages. For example, tectonic movement usually results in large changes in the land and water constitution of an area. Since land and water react to solar radiation in different ways, these new arrangements may lead to significant increases in snow and ice in some areas.

Another possibility arises because tectonic movement often involves the submersion of large areas of land, reducing rocky and vegetative materials that may release or absorb carbon dioxide, a component of the atmosphere crucial to determination of Earth's average annual temperature. The collision of tectonic plates frequently results also in the rise of volcanic activity, in which carbon dioxide and other gases are released to the atmosphere. Volcanic eruptions have been clearly found to have had at least some climatic effects on modern times, although not necessarily of the size needed to produce a glacial effect. In general, most researchers seem to agree that tectonic movement would probably have been a more important factor in climate change early in Earth's history than in more recent millennia (DeConto 2009).

Another possible natural cause of climate change is solar phenomena. The Sun is an active astronomical body undergoing changes, both large and small, all the time. One of the best known of these changes is the appearance of sunspots, areas on the Sun's surface where temperatures are significantly less than areas around the spots. Sunspots have been known to human observers since they were first described in 1128 by an English monk, John of Worcester. They were not well studied until the early 17th century, when they were first seen by telescope by English astronomer Thomas Harriot and Frisian astronomer brothers Johannes and David Fabricus.

The occurrence of solar phenomena, such as sunspots and flares, means that solar insolation undergoes regular fluctuations that might be associated with climate changes on Earth. One of the most famous connections ever made by some researchers is a period called the Maunder Minimum, which occurred between about 1645 and 1715. Observations made at the time found an unusually low number of sunspots and, hence, reduced solar insolation. The important point about that observation is that it corresponds with the period discussed earlier known as the Little Ice Age. Some observers use this connection to confirm the fact that solar phenomena can have a significant effect on Earth's climate (Shindell et al. 2001).

Critics have raised questions about the solar phenomena/climate change connection. For one thing, scientific data about sunspots, flares, and other solar events date back only a short period of time, about 400 years at best. No data of any kind exists to connect solar activity with climate back to the thousands or millions of years over which climate change is often discussed. It appears that most climatologists who write about the topic take a somewhat in-between position, agreeing that solar phenomena are, indeed, involved in climate change, but that the magnitude of that effect is small compared to other factors (Hegerl et al. 2007; Ogurtsov et al. 2015).

A Greenhouse Earth

Prior to the 19th century, almost no one would have thought about the contribution that human activities made, if any, to climate change. The cobbler working in his small shop; the wife mending sheets at the kitchen table; the farmer plowing his fields. How could such mundane activities affect changes in Earth's temperature, precipitation, storm patterns, and other climate events?

That line of thought began to change in the late 18th century because of two events: the rise of the Industrial Revolution and advances in meteorological research. The Industrial Revolution has been defined as the period in human history when work traditionally done by humans and other animals was being taken over by machines. Some examples of that machinery are the first machine to pump water out of mines (Thomas Newcomen, 1712), the first factory (John Lombe, 1721), the flying shuttle (John Kay, 1733), the spinning jenny (James Hargreaves, 1764), the Watt steam engine (James Watt, 1775), the first method for making wrought iron (Henry Cort, 1784), the cotton gin (Eli Whitney, 1794), and the steam locomotive (Richard Trevithick, 1804).

One common feature of this transfer of production methods was the role of fossil fuels. Fossil fuels are combustible substances formed millions of years ago and now buried (usually) deep within the Earth. They include coal, oil, and natural gas. Almost all of the new machines developed after 1700 operated by means of the combustion of a fossil fuel. At first, wood was the fuel of choice. It was readily available and fairly inexpensive. But the demand for wood soon grew to a point in the late decades of the 17th century that virtually all forests in Europe had been depleted (Kaplan, Krumhardt, and Zimmermann 2009).

The connection between the Industrial Revolution and climate change can be expressed in one simple equation and one of its derivatives:

$$C + O_2 \rightarrow CO_2$$

$$(CH_2O)_x + O_2 \rightarrow xCO_2 + xH_2O$$

All fossil fuels contain carbon as a major component. The carbon may occur uncombined, as an element (C), as in some kinds of coal. Or it can occur as a combustible compound, made of carbon, hydrogen, and/or oxygen, along with other elements, such as sulfur and nitrogen. In either form, combustion results in the conversion of carbon to carbon dioxide (CO_2) and hydrogen, if present, to water (H_2O). Some carbon is also converted to another oxide of carbon, carbon monoxide (CO). Other elements present in small amounts in the fuel are also converted to their oxides: sulfur to sulfur dioxide or sulfur trioxide, and nitrogen to any one of five nitrogen oxides.

And this is the point at which meteorological research becomes relevant. The release of carbon dioxide to the atmosphere turns out to have very significant effects on the atmosphere's composition and that, in turn, has a strong influence on weather and climate.

The amount of carbon dioxide in the atmosphere changes over time. Prior to the Industrial Revolution, the concentration of carbon dioxide in the atmosphere was thought to be about 0.028 percent of the atmosphere. It is thought to have varied from about 0.018 percent during the ice ages to 0.028 during interglacial periods ("CO_2 at NOAA's Mauna Loa Observatory Reaches New Milestone: Tops 400 ppm" 2013). That amount of carbon dioxide comes from natural sources such as the decomposition of organic material; respiration of plants, animals, and soil; and release of the gas from oceans.

Atmospheric carbon dioxide is an essential component of the atmosphere because of its role in the maintenance of life on Earth. As almost any high school student knows, the basis of life on Earth is the process of photosynthesis, the process by which carbon dioxide and water are converted to the carbohydrates of which plants are made. Plants are the primary food on which animal life on Earth survives. Photosynthesis also results in the formation of, and is the main source of, oxygen, which almost all living organisms need for their survival. Under most circumstances a natural balance exists in the amount of carbon dioxide in the atmosphere and fixed in living organisms on Earth. The amount of the gas released to the atmosphere by respiration, decomposition, forest fires, volcanic eruptions, and other events is roughly equal to the amount absorbed by the oceans and used up in the process of photosynthesis.

Most of the information about carbon dioxide cited here is of relatively recent origin in the history of science. The compound was not even discovered until 1753, when Scottish chemist Joseph Black prepared the gas in his laboratory by adding acids to calcium and magnesium carbonate. He continued to study the new compound and a few years later was able to prove that carbon dioxide exists in the atmosphere. Its role in determining Earth's climate was not discovered until about a century later.

Breakthroughs in Global Warming Research

The first breakthrough in that discovery came in the mid-1820s as a result of the research of French mathematician Jean-Baptiste Fourier. Fourier was intrigued by a simple problem with regard to Earth's climate. He was able to calculate the amount of heat that one should expect on Earth, given the size of the planet and its distance from the Sun. But those calculations failed to fit the data. That is, given the existing astronomical relationship between the Sun and the Earth, our planet should be much cooler than it actually is. Where does the "extra" heat come from, Fourier asked.

A partial answer to that question came when Fourier recognized that Earth is heated when *visible sunlight* strikes Earth's surface. Some of that sunlight is reflected in the form of *infrared radiation*. For some reason, he discovered, the infrared radiation did not penetrate Earth's atmosphere as efficiently as sunlight did in reaching the atmosphere. The atmosphere ended up retaining the extra heat that was unable to escape back into space. Fourier decided that Earth must be surrounded by some sort of "blanket" that prevented heat from escaping Earth once it had been warmed by the Sun. He had no idea what that "blanket" might be, but he compared it to a piece of equipment that was familiar to researchers of the time. It was a box made of solid wood on all four sides and the bottom, covered with a transparent material across the top. When the box was exposed to sunlight, the interior of the box began to warm up. To anyone looking at such a box today, a comparison could be made with the kind of greenhouse in which plants are grown. The discovery Fourier made, then, became known as the *greenhouse effect*, although he himself is not known ever to have used the term.

The next chapter in this story of climate science is usually a mention of the research of English physicist John Tyndall in the mid-19th century. This research consisted of a series of experiments on the ability of various gases to trap heat. He found that carbon dioxide was among the most efficient of gases in doing so and hypothesized that the ice ages might have come about at least in part because of a decrease of carbon dioxide in Earth's atmosphere (Black 2011).

The problem with this part of the story is that another researcher had already made such a discovery. In 1856, the eminent American scientist and Secretary of the Smithsonian Institution Joseph Henry read a paper, "Circumstances Affecting the Heat of the Sun's Rays," at a meeting of the American Association for the Advancement of Science. The paper described experiments similar to those reported by Tyndall five years later. The author of the paper, however, was not Henry himself, but Eunice Newton Foote. We don't know much about Foote, including her education in science. But her paper was well written, describing experiments that were carefully designed and rigorously conducted. The reason that her research was never widely disseminated or recognized in the scientific record is not clear. But its place in the history of climate science is certainly deserving of mention (McNeill 2016).

Ignoring for the moment who it is that should receive credit for this discovery, the fact seems to be that Tyndall pursued the consequences of his research to a greater extent than any other scientist of the time. The most important simple idea he developed is that the "blanket" to which Fourier had referred was likely to be a gas or gases in the atmosphere that kept heat from Earth's surface from escaping back into space. At one point, he wrote:

> Aqueous vapour is a blanket, more necessary to the vegetable life of England than clothing is to man. Remove for a single summer-night the aqueous vapour from the air which overspreads this country, and you would assuredly destroy every plant capable of being destroyed by a freezing temperature. (Tyndall 1869, 417)

Tyndall's research and his ideas on atmospheric warming influenced several of his peers and followers. In 1896, for example, Swedish chemist and physicist Svante Arrhenius began a study of the effects of carbon dioxide, rather than water vapor, on Earth's climate. (Arrhenius had previously studied the dissociation of acids and bases, research for which he was awarded the Nobel Prize in Chemistry in 1903.) He calculated that cutting the amount of carbon dioxide in the air by half would produce sufficient cooling to cause an ice age on the planet. Arrhenius also developed a second idea of considerable importance to climate science: positive feedback. The term *positive feedback* refers to any event in which the product of some action results in an increase in the likelihood that that reaction will re-occur with even greater effect. An example of positive feedback in climate science noted by Arrhenius is the melting of glaciers and ice fields by increasing the amount of carbon dioxide in the atmosphere. As carbon dioxide levels in the atmosphere rise, more heat is retained in the atmosphere. That additional heat causes glaciers and ice fields to begin to melt. As melting occurs, areas that were originally covered by a white material (snow) are now covered by a dark material (soil). Dark materials absorb heat more readily than do white materials. So, the melting of glaciers and ice fields tend to produce conditions under which more heat is retained by Earth's surface. (Arrhenius's classic article on this topic is available online at Arrhenius 1896, especially, page 267 et seq.)

Arrhenius's results were received with something less than enthusiasm by many of his colleagues. His greatest critic at the time was Swedish physicist Knut Ångström. Ångström decided to conduct his own experiments on the absorption of infrared radiation by various gases. He had an assistant construct a tube that could be filled with different gases and then exposed to infrared radiation. He found that carbon dioxide absorbed relatively little radiation, certainly not enough to explain temperature drops characteristic of the ice ages. As it turned out, Ångström's experiments contained some fundamental errors that invalidated his results. For example, his finding that reduction of carbon dioxide in his tubes produced a decrease of only 0.4°C was off by a factor of 250 percent, an error to which Arrhenius himself pointed in an article criticizing Ångström's work. As it developed, most researchers of the time accepted Ångström's conclusions, rather than those of Arrhenius (Mason 2012).

Anthropogenic Effects on Global Warming

Fourier, Tyndall, Arrhenius, Ångström, and their colleagues were primarily interested in theoretical musings about ways in which atmospheric gases might produce changes in Earth's temperature. They paid relatively little attention to the ways in which their research might provide information about the way these changes might affect life on Earth. Arrhenius himself, in an 1896 lecture, acknowledged that one conclusion of his research was that an accumulation of carbon dioxide in the atmosphere from industrial sources might eventually produce measurable and significant effect on the planet's temperature. But he thought that such changes were thousands of years in the future. More to the point, he suggested that humans today "would then have some right to indulge in the pleasant belief that our descendants, albeit after many generations, might live under a milder sky and in less barren surroundings than is our lot at present" (Sample 2005).

Other researchers took this feature of global warming somewhat more seriously, and they began to look for specific ways in which climate changes might affect the planet and its inhabitants. The key question in these studies was not what the effect of carbon dioxide has on global temperatures but how carbon-dioxide concentrations in the atmosphere could change as the result of human activities. Notable among these individuals was American geologist Thomas C. Chamberlin. In referring to Arrhenius's research, Chamberlin pointed out that Arrhenius "does not, however, postulate the conditions which control the enrichment and depletion of the atmosphere which has been the essential endeavor of this paper." The goal of that paper, he goes on, has been "to explain the profound glacial oscillations" (Chamberlin 1908).

The explanation that Chamberlin gives for this phenomenon is that Earth consists of a complex system of agencies that provide carbon dioxide to and remove it from the atmosphere. He mentions the role of the oceans as a sink for carbon dioxide, as well as an important source for its return to the atmosphere. He also points to other natural factors, such as volcanoes and the presence of carbonate rocks as sources and sinks of carbon dioxide. Indeed, his argument is a precursor of the elements of what we refer to today as the carbon cycle, of which atmospheric carbon dioxide is only one part. Chamberlin claims that this system has and continues to remain in a "delicate balance" that is "congenial to life." Probably the most important feature of Chamberlin's work was his anticipation of mathematical and physical models that are so widely used to predict future climate changes (Weart 2018).

Another researcher to whom Arrhenius himself refers at great length in his paper "On the Influence of Carbonic Acid in the Air upon the Temperature of the Ground" (1896) was Swedish geologist Arvid Gustaf Högbom. Arrhenius actually includes a five-page translation of Högbom's research in his own paper. Högbom had conducted research to measure the amount of carbon dioxide in the atmosphere and on land and in the oceans. In his studies, he included estimates of the amount of carbon dioxide released by industrial activities. He found that that quantity was roughly equal to all the carbon dioxide released from natural sources (such as evaporation from the oceans) and only about a thousandth of all carbon dioxide already resident in the atmosphere. The interesting point Högbom made, however, was that that number, although insignificant at the time, could become more important if industrialization increased over time. His comment was, of course, prescient of the changes that were actually to occur over the next century (Arrhenius 1896, 269–273).

So much of our understanding about climate change depends, perhaps obviously, on the technology available for making precise measurements. One of the earliest examples of that truism occurred in 1938, when English steam engineer Guy Stewart Callendar published an article "The Artificial Production of Carbon Dioxide and Its Influence on Temperature" in the *Quarterly Journal of the Royal Meteorological Society*. In addition to his regular profession, Callendar was fascinated by problems of weather and climate. His paper reported on a study he conducted of temperature measurements dating back a half century from 147 meteorological stations around the world. He calculated that Earth's temperature had been increasing by 0.003°C per year. He attributed this increase to the release of an estimated 150 billion tons of carbon dioxide over the study period, three-quarters of which he said had remained in the atmosphere. Callendar's paper was the first rigorous analysis of the connection between carbon dioxide from industrial sources and changes in the planet's annual average temperature (Callendar 1938).

A New Day for Climate Science

People alive in the third decade of the 21st century might be excused for being surprised at the limited role that science, especially basic science, played in the everyday lives of men and women prior to World War II. One classic example of this fact dates to the beginning of World War I. At the outset of that conflict, the American Chemical Society offered to provide chemists to the U.S. War Department to assist in the war effort. The War Department's response was that it had checked its personnel records, and the department responded, "Thanks, but we already have a chemist" (Dittmann 1987).

Historically, basic research had been essentially a subject of little interest to governmental agencies and the general public. It has drawn the attention and financial support of corporations largely because of the practical applications such research might produce (see, e.g., de Solla Price [1963] 1986). That story holds true for much of the research on climate and climate change. As the previous discussion has illustrated, many important researchers in the field were not necessarily meteorologists, but specialists in physics, engineering, or some other field of endeavor.

A major exception to this pattern occurred during wars. At that point, any useful information about explosives, weather patterns, ocean currents, or other phenomena that might be of value in military operations was encouraged. Indeed, the ultimate victory of the Allied nations during World War II has often been credited to the development of a new long-distance ranging technology, radar (Douglas 2018).

World War II research was directly responsible for a number of developments in climate science shortly after the end of the war. For example, studies of the structure and composition of

Earth's atmosphere provided far more accurate and detailed information about the role of carbon dioxide in heating and cooling than had ever been available from previous, Earth-bound experiments ("How to Talk to an Ostrich: 'Who Says CO_2 Heats Things Up?" 2012). As one historian has written, "The Second World War and its aftermath brought a phenomenal increase in observations from ground level up to the stratosphere, which finally revealed all the main features." Research also confirmed and extended information about the role played by carbon dioxide in the ocean, as well as other ocean features, which were later shown to be even more accurate than post-war research (Weart 2019b). It was the availability of this research following the war that made possible advanced studies of the role of carbon dioxide and other gases found in the atmosphere and in the oceans that produced an acceleration of climate research in the 1950s.

Climate Models

Data collected from numerous observations of atmosphere, oceans, and other parts of the Earth provided climate researchers with a huge opportunity to learn more about changes in weather and climate. One problem they soon realized, however, was the pure physical and mathematical challenge of using these data. As one historian has written, "the climate system is too complex for the human brain to grasp with simple insight. No scientist [prior to World War II] [had] managed to devise a page of equations that explained the global atmosphere's operations" (Weart 2019b). Fortunately, a solution to this problem was at hand, one that had also been developed during and for conduct of the war: the computer. Researchers began to explore the possibility of providing this mechanical device with a host of observations about the atmosphere, or some other component of climate change, and letting the computer predict the weather and/or climate that could be expected based on the data supplied.

The first successful device of this kind was invented by American meteorologist Norman A. Phillips in 1956. Phillips supplied a very simple type of computer with a set of equations that described the behavior of various parts of the atmosphere, along with observational data available for each variable in the equation. The computer was then asked to predict climatic conditions at some given time over some given area. In the early stages, the concept was tested by asking the computer to predict existing or known climate patterns for a simplified model of Earth over a limited period of time. The computer successfully completed this assignment, and the day of computer modeling of climate had arrived ("Climate Models" n.d.; Weart 2019b).

The primary goal of Phillips's research was not, of course, to "predict" the existing climate; we already knew what that was. Instead, the model was designed to predict what the climate might be like in 100 years, 1,000 years, or at some other future time. The process involved giving a computer not data for existing conditions, but for possible conditions at some time in the future. The question asked the computer might be, for example, suppose the concentration of carbon dioxide reaches 450 parts per million (ppm) at some time in the future. (It is currently just above 400 ppm.) What will the climate be like under these circumstances? (Of course, the variety of data one can provide a computer is very large, so very sophisticated answers can be produced.)

The model developed by Phillips is called a *general circulation model* or *global climate model* (GCM, in either case). As the name suggests, such models attempt to predict future climate conditions for Earth as a whole. Other types of models are also possible, such as regional climate models that describe conditions in only a certain part of the planet (Goosse et al. 2008–2010).

One of the earliest discoveries about future climates resulting from computer modeling was that reported by American physicist Gilbert N. Plass. (Like many of his predecessors, Plass took on the task simply out of curiosity, unconnected to his regular work on weapons engineering.) Plass confirmed some earlier findings that doubling the concentration of carbon dioxide in the atmosphere would bring about an increase in global temperatures of about 3°–4°C. He also reported that continuing increases of carbon dioxide from anthropogenic sources would result in an increase of global temperatures in the range of about 1.1°C per century (Plass 1956). As with many earlier climate science predictions, Plass's report was largely ignored by his colleagues and the general public (Weart 2017).

New Discoveries

More powerful tools for analysis of data (GCMs) were of value only to the extent that more and better data about climatic conditions became available. The last few years of the 1950s saw these data begin to appear. One of the first reports of special significance arose out of the research of American oceanographer Roger Revelle. Among his many research projects on the composition of seawater, Revelle explored the tendency of the oceans to act as a sink for carbon dioxide. The issue was of some importance because many earlier scientists had taken the position that accumulation of the gas in the atmosphere was not much of a problem because of its tendency to dissolve in seawater. So what if humans dumped larger amounts of carbon dioxide into the air, they said. That extra carbon dioxide would just dissolve in the ocean and not have an effect on Earth's climate.

Revelle's research revealed a serious flaw in that argument. While it was true that most of the carbon dioxide released to the atmosphere would eventually dissolve in ocean water, much of it would then evaporate back into the atmosphere again. And, Revelle pointed out, the increasing production of anthropogenic carbon dioxide hinted that the gas would continue to accumulate in the atmosphere, accompanied by an increase in global temperature. Revelle and his colleague Hans E. Suess published a summary of their findings on this topic in a now-famous paper "Carbon Dioxide Exchange between Atmosphere and Ocean and the Question of an Increase of

Atmospheric CO, during the Past Decades" in 1957 (Revelle and Suess 1957).

Revelle was seriously disturbed by his findings and, like some of his climate predecessors, issued dark warnings about a possible future produced by a carbon-dioxide-laden atmosphere. Both in his own writings and in reports of his work in general newspapers, terms such as "global warming" and "climate change" began to appear, if not for the first time, at least more and more frequently. It had become time for the general public to wake up to the threats of these two phenomena (Weart 2007; the frequent mention of Spencer Weart's research in this segment is a testimony to the essential nature of this long, detailed, and easily understandable history of climate change developments in the last decades of the 20th century).

If there is any single bit of research that is well known and influential in the late decades of the 1950s, it is the work of American chemist Charles David Keeling. Keeling had earned his bachelor's degree at the University of Illinois in 1948 and his doctorate in chemistry at Northwestern University in 1954. He chose to do his postdoctoral research at the California Institute of Technology, after which he joined the Scripps Institute of Oceanography in 1956. He retained his affiliation at Scripps until his death in 2005.

At Scripps, Keeling learned about and became fascinated with Roger Revelle's research on carbon dioxide in the atmosphere. He became convinced that additional, more precise measurements of carbon dioxide in the atmosphere were needed to drive climate change theory. He started out his own research in the area by setting up simple research stations at a number of locations, from Big Sur, California, to the Olympic peninsula, in Washington state. He was intrigued to discover that he obtained similar results at all locations. Carbon dioxide ratios reached a maximum in late afternoon, then dropped off during evening hours. He wondered how this change could occur.

An important problem with Keeling's research, however, was one familiar to other climate researchers of the time:

noise-contaminated data. That is, carbon dioxide is ubiquitous in the environment, but somewhat variable because of local conditions (forests versus industrial areas, as an example). The common belief among experts in the field at the time was that it would be necessary to collect carbon-dioxide data over many years from many different locations across the planet to get a reliable idea as to ambient carbon-dioxide concentrations (Rosby 1959, 15). Keeling, however, had a different idea. Why not establish an observatory, he thought, in some location where atmospheric conditions were natural and constant. That location would have to be at some distance from the usual sources of anthropogenic carbon dioxide (such as factories and major highways), and even uncontaminated by natural conditions (the growth and death of plants, for example). Such an observatory could be used to determine carbon-dioxide concentrations in as pure an atmosphere as possible.

The choice Keeling made was the creation of an atmospheric laboratory on top of Mauna Loa, Mauna Loa is the highest point in the Hawaiian Islands, with its top extending above the inversion layer that would otherwise hamper "pure" measurements of carbon dioxide. The site chosen for the new observatory had already been minimally developed by the U.S. military in World War II, with an undeveloped road leading from the mountain's base to its summit. Keeling began his observations in 1958. Only two years later, he felt that he was able to write a paper on his discoveries to that point. In that paper, Keeling produced a graph showing carbon-dioxide measurements from four different sources, one of which was the Mauna Loa site. The data from that location had already begun to take the shape of what was eventually to become known as the Keeling Curve. It showed a maximum concentration of carbon dioxide of about 320 ppm in both years, with minima of about 310 ppm.

Keeling continued his observations throughout the remainder of his professional career, almost until his death in 2005. You can see his earliest results (1958–1959) at Keeling (1960). The National Oceanic and Atmospheric Administration

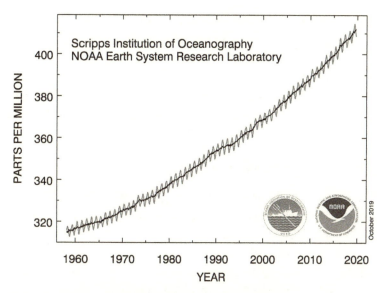

Figure 1.1   Atmospheric CO$_2$ at Mauna Loa Observatory

Earth System Research Laboratory, Global Monitoring Division, National Oceanic and Atmospheric Administration. https://www.esrl.noaa.gov/gmd/ccgg/trends/. Updated October 7, 2019.

updates the graph regularly, showing results over more than a half century of observations (figure 1.1).

Two features of the Keeling curve require special comment. First, notice that the curve is not a continuous line but a repetitious up-and-down ("sawtooth") line with an overall increase in carbon-dioxide concentration over time. Why the "sawtooth" pattern? Easily explained. In the fall, plants and other organisms die off, releasing carbon dioxide to the atmosphere. In the spring, plants begin to grow, taking carbon dioxide out of the atmosphere in the process of photosynthesis; carbon dioxide is removed from the atmosphere.

Second, the really significant aspect of Keeling's research is apparent when placed in the context of changes in carbon-dioxide concentrations and global temperatures over the past two millennia ("1700 Years of Global Temperature Change from Proxy

Data" 2017). These data show that after relatively moderate fluctuations in temperature between about 300 C.E. and 1900 C.E. temperatures began to increase quite dramatically. While these data are often taken as strong evidence of anthropogenic effects on global warming and climate change, some critics have offered objections to this conclusion from the Keeling data (Mann 2015; McKitrick, n.d.; Muller 2004).

Post-Keeling Developments

The research of Charles Keeling might be thought of as a turning point in the history of climate change science. Prior to these studies, climate change research was a somewhat esoteric subject. As noted earlier, individuals interested in climate often conducted their research as a hobby or side -interest to their major field of work. No field of "climatology" could really be said to have existed, and there were few scientists who might list themselves as "climatologists" (Weart 2019a). That situation began to change only at about the time that Keeling was conducting his research. The first textbook in climatology, for example, was not published until 1965 (Sellers 1965) and the first journal on the topic appeared only in 1977 (Barry 2013). Even today, students interested in majoring in climate studies generally find themselves in departments of geophysical sciences, environmental sciences, meteorology, or some other related topic.

So Keeling's research was part of a new movement among a handful of scientists to learn more about Earth's climate, about changes taking place in the climate, and about reasons for such changes' occurrence. Specialists in the field soon discovered that they needed not only to develop new and improved technology with which to collect new and better data but also to develop organizations through which they could disseminate their findings. They also saw the need to alert the general public and policy makers about their findings and their potential impact on human civilization.

Scientific Developments and the Responses of Scientists
Research Developments

By the early 1960s, many of the basic techniques of climate proxies had been invented and developed and were in relatively wide use. As an example, ice core technology was being extended to greater and greater depths, providing increasingly more extensive and detailed information about primitive climates. The greatest depth to which cores had reached as of 2020 was 3,623 meters (11,886 feet), achieved at the Vostok research station in East Antarctica in January 1998. Although ice coring has been carried out on all of Earth's continents, the greatest fraction by far have been conducted in Greenland and Antarctica. (For an especially interesting review of what is probably the world's most famous ice core station—Vostok—see Ueda and Talalay 2007.)

Some new climate change proxies and variations of existing proxies also began to appear in the post-Keeling years. One such proxy can be traced to a revelation of American chemist Harold Urey in the late 1950s. Urey was an authority on the properties and applications of radioactive isotopes. At one point, he is said to have been examining the oxygen isotopic composition of fossil mollusks and exclaimed, "I suddenly found myself with a geologic thermometer in my hands" (Emiliani 1958, 54). Urey was not convinced of the practical value of this discovery, however, and he assigned the project to one of his graduate students, Cesare Emiliani, for further study. Eventually, Emiliani was able to develop a method by which the ratio of oxygen isotopes in fossil foraminifera can be used to estimate climatic conditions in the distant past (Pearson 2012).

A somewhat similar breakthrough occurred in the work of American geochemist Wallace Broecker in the 1950s. Broecker was intrigued by strata of fossil coral reefs in many parts of the world. He became convinced that the composition of those corals could provide information about climatic conditions dating back hundreds of thousands of years. Broecker, like

Emiliani, also took advantage of the newly developed science of radioactive isotope dating to calculate the ages and related climatic conditions of corals farther back than had ever been possible before. Instead of carbon- or oxygen-dating, however, he used uranium isotope dating, valid (as described earlier) for much longer periods of time in history. The greatest value of Broecker's research was to confirm the earlier theories of Milutin Milanković about the astronomical basis of long-term climate change and to provide descriptions of smaller segments of the period's described in Milanković's theories (Broecker 2002).

Aided by the rapid development of computer technology in the 1950s and 1960s, the science of climate modeling also began to develop at a comparable rate. Perhaps the most frequently mentioned names in the field at the time, and since, were those of Japanese-American meteorologist and climatologist Syukuro Manabe and American meteorologist Richard T. Wetherald. In 1967, Manabe and Wetherald published a paper "Thermal Equilibrium of the Atmosphere with a Given Distribution of Relative Humidity" that, for the first time, represented all of the basic elements of Earth's climate in a mathematical model, from which predictions about future climate changes could be made. In one poll of climatologists, the paper has been voted "the most influential climate change paper of all time." One participant in that poll described the Manabe-Wetherald paper as "the first physically sound climate model allowing accurate predictions of climate change" (Pidcock 2015).

The specific contribution of greatest importance in the Manabe-Wetherald paper was an effort to model the exchange of energy between Earth's surface and the troposphere, and then to calculate changes that would occur if additional carbon dioxide was added to or removed from the atmosphere. The authors found that doubling the amount of carbon dioxide in the atmosphere would produce an increase in global temperatures of about 2.4°C (4.3°F).

Computer models have continued to develop since the classic 1967 paper. For example, Manabe was involved again in a classic experiment completed in 1969 in which a more extensive group of variables was included in the model. The model was called a coupled atmosphere-ocean general circulation model, in which the influence of oceans, ice, and other parts of the hydrosphere were added to the previous atmosphere-surface model. Another extraordinary step forward, the model was hampered to some extent by its complexity. The first run of the model took 46 hours to complete.

Climate modeling is now a popular field of climate research with programs located in academic and governmental institutions around the world. In the United States alone, for example, the federal government conducts modeling programs at the National Center for Atmospheric Research (NCAR), Geophysical Fluid Dynamics Laboratory (GFDL), Goddard Institute for Space Studies (GISS), National Center for Environmental Modeling (NCEP), Goddard Global Modeling and Assimilation Office (GMAO), and Center for Multiscale Modeling and Prediction (CMMAP) ("Developers of Climate Models" 2012; this website advises that this list is not a complete list of all climate modeling activities, but, instead, only a sample of some of the larger efforts). (For an excellent review of the history of climate modeling, see Hickman 2018; good general introductions to climate modeling can be found at Crimmins, n.d., which draws on many primary sources; Goosse et al. 2008–2010; Stocker 2016.)

Scientists React

For many members of the scientific profession, the flow of new and better data about global warming and climate change had a single critical message for the world: Earth's climate had begun to alter significantly as a result of the Industrial Revolution. That change had continued through the modern era, and it held troubling consequences for the planet's future. That viewpoint was, however, not the unanimous verdict of professionals in the fields of climatology and related sciences. Many authorities held less extreme positions, ranging from total rejection of the theories of Milutin Milanković, to the acceptance of those theories for past history but doubts about current trends, to acceptance of changing temperatures and carbon dioxide in the atmosphere but rejection of or doubts about the role of human activities in climate change. The debate among experts in (and around) the field of climatology was not a binary dispute, "yes" or "no" about climate change, but a host of in-between positions about Earth's current and future climate prospects. For example, Cesare Emiliani, in writing about his discoveries in the journal *Science* in 1966 predicted that "a new glaciation will begin within a few thousand years" (Emiliani 1966). Only six years later, however, Emiliani was somewhat more cautious in his views. He pointed out that Earth's climate was a system of delicate balances, and that global warming was at least as likely as global cooling. What this means, he said, is that "Man's activity may either precipitate this new ice age or lead to substantial or even total melting of the ice caps" (Emiliani 1972).

Individuals who were dubious about the consensus view of global warming and climate change in the post-Keeling period sometimes pointed to an alternative theory of climate change: global cooling. The theory that Earth might be experiencing (and would continue to experience) *cooler*, rather than *warmer*, temperatures had a long history in science. This theory was often phrased in terms of a return of the planet to a "new ice age," with predictions as to when and how that era might appear. These commentaries were often based on weather records over a relatively short period of time, such as a few decades. Concerns about a global cooling reached a maximum, however, in the 1970s, often as the result of data collected a decade or two earlier. These data seemed to suggest that a period of planet-wide cooling had begun, a pattern that some observers thought might presage the arrival of a new ice age on Earth (see, e.g., Budyko 1969; Starr and Oort 1973; the graphs on which these theories were based can be found at Richard 2017). These science articles often led to somewhat hysterical articles in local newspapers about the coming age of glaciers (see, e.g., Watts 2013b).

Other climate change skeptics, while generally willing to accept some level of global warming in coming years, suggested that the change was probably not one of much significance. In 1970, for example, American climatologist Helmut E. Landsberg wrote an article in which he attempted to "sort fact from fiction" about global warming. He concluded that, although such a process was likely to occur, such a change could "hardly be called cataclysmic" (Landsberg 1970).

This range of views about global warming and climate change has continued over the past half century, with some writers firmly committed to the belief in a warming Earth, threatened by human activities to some potentially disastrous consequences, to questions about the basic assumptions and data dealing with Earth's climate, and suggestions that warnings of forthcoming catastrophes are overblown and even outright false. (As examples of more recent expressions of those skeptical views of climate change, see Soon and Baliunas 2003; Yarbrough 2015.) A more detailed discussion of climate change skeptics is available in Chapter 2 of this book.

In many ways, "climate science" in the 1950s, 1960s, and 1970s was not very different from the "climate science" of earlier history. There was, in fact, no such discipline as "climate science" during those decades. Instead, experts from a variety of fields—such as meteorology, oceanography, physics, chemistry, agriculture, and atmospheric sciences—with interest in climate were carrying out their own research on climate issues. The relative handful of experts working in the field of "climate science" soon found ways to join together to share mutual research interests and findings.

Among the first of those sessions was a meeting held under the auspices of The Conservation Foundation, an organization founded in 1948 with the goal of protecting the world's natural resources. (It later evolved into the modern-day World Wildlife Foundation.) The meeting consisted of a group of chemists, physicists, ecologists, and other scientists "with particular experience and interest in the [climate] problem" ("Implications of Rising Carbon Dioxide Content of the Atmosphere" 1963, i). In their discussions, meeting participants reviewed existing knowledge about carbon dioxide and its effects on global temperature. They concluded that the most troubling point about the whole issue was that so little was known about carbon dioxide and its role in the environment. They recommended an aggressive plan for increasing research on the compound and its effects on Earth systems. In conclusion, participants also noted the importance of educating the general public about future scenarios involving climate change. "It is very important," the final report said, "to alert more people, more scientists and more scholars in the social sciences as well as the pragmatical sciences, to the need for planning and the realization that there is an obligation to provide for the future as well as the present" ("Implications of Rising Carbon Dioxide Content of the Atmosphere" 1963, 15)

Among other important meetings and actions to occur over the next half century were the following:

> 1979: The World Meteorological Organization issued a call for the First World Climate Congress, in Geneva, Switzerland. The conference was attended by more than 350 experts in the field of climate science from 53 countries and 24 organizations. Attendees came from a wide variety of specialties, including agriculture, water resources, fisheries, energy, environment, ecology, biology, medicine, sociology, and economics. A declaration issued at the end of the conference called for, among other things, a program
>
> "(a) To take full advantage of man's present knowledge of climate;
>
> (b) To take steps to improve significantly that knowledge;
>
> (c) To foresee and prevent potential man-made changes in climate that might be adverse to the well-being of humanity." ("A History of Climate Activities" 2009)

The conference also recommended the creation of a World Climate Programme, with the purposes of improving the quality of data about climate change and aiding individuals and organizations in making the best possible use of these data. That program continues its operations today ("World Climate Programme" 2019).

1985: One of the most comprehensive of all early climate change conferences was sponsored by the United Nations Environment Programme, the World Meteorological Organization, and the International Council for Science. It was held in Villach, Austria, to review current knowledge about carbon dioxide in the atmosphere and its effects on global climate. The conference consisted of about 80 scientists from 29 different countries. Its final report not only summarized the scientific knowledge reviewed at the meeting but also recommended the creation of an international body to continue research on climate change and prepare recommendations for dealing with the growing problem ("Report of the International Conference on the Assessment of the Role of Carbon Dioxide and of Other Greenhouse Gases in Climate Variations and Associated Impacts" 1986).

1988: As a follow-up on the recommendations of the Villach conference (1985), the United Nations Environment Programme and the World Meteorological Organization combined forces to create the Intergovernmental Panel on Climate Change (IPCC). The organization was created to provide the world's governments with regular updates on the status of climate research, along with recommendations for actions dealing with global warming and other changes. As of 2020, the organization has issued five reports, in 1990, 1996, 2001, 2007, and 2014. Its next report is scheduled for releases in 2022 ("Reports" 2019). Each report consists of several chapters dealing with technical information on a number of features of climate change, along with a summary for policy makers, based on existing information. IPCC reports are almost certainly the single most valuable resource for information on global climate change available ("Preparing Reports" 2019).

1992: In preparation for the Earth Summit conference, to be held in Rio de Janeiro, Brazil, in June 1992, an Intergovernmental Negotiating Committee met to prepare a document on climate change to include in the Rio conference. The document, the United Nations Framework Convention on Climate Change, was ultimately adopted at the Rio meeting and signed by 165 nations. It was later ratified by the 193 members of the United Nations plus the State of Palestine, Niue, Cook Islands, and the European Union. The purpose of the treaty was to achieve "stabilization of greenhouse gas concentrations in the atmosphere at a level that would prevent dangerous anthropogenic interference with the climate system" ("United Nations Framework Convention on Climate Change" 1992, 4) As provided for in the treaty, parties to the convention have continued to meet annually to assess progress toward this goal. The most important of those annual meetings was held in Kyoto, Japan, in 1997.

1997: The Kyoto meeting in 1997 was historic and contentious. One of the critical issues, differences in the way developed and developing nations were treated in setting carbon-dioxide limits, was resolved only in the last few hours prior to adjournment of the meeting. Eventually, a document was agreed upon that acknowledged the significance of level of development in a nation's contribution to global warming. The United States signed the treaty, but five years later, President George W. Bush withdrew from the agreement. The United States has not been a party since that time, although Kyoto partners have proceeded with refining and expanding that document in succeeding years ("UNFCCC Process-and-meetings" 2019).

Responses from the Body Politic

As the previous section suggests, members of the scientific community gradually gravitated to an acceptance of the fact that planetary warming had begun to occur to a significant degree and that human activity was probably the most important cause for that change. The pattern at the time seemed to be that the more closely one's research involved climate studies, the more likely he or she was convinced that change was occurring and that anthropogenic causes were involved. By the early 2000s, one survey of 928 papers on climate change found that "none of the papers disagreed with the consensus position [on anthropogenic causes of climate change]" (Oreskes 2004).

The views of policy makers and the general public tended to lag behind this consensus. Public opinion polls well into the 21st century continued to find a significant fraction of the population in each case that did not believe that climate change was taking place, that human activities were responsible for any changes that was occurring, and/or that climate change represented a threat to the future of human civilization. Indeed, as late as 2018, a survey conducted by the Pew Research Center found that, while 83 percent of Democrats and Leaning Democrats in the United States saw climate change as a "major threat" to the country, only 27 percent of Republicans and Leaning Republicans expressed the same view (Poushter and Huang 2019, 10). That pattern changed somewhat dramatically among the former group between 2013 and 2018, increasing by 25 percent, but has largely not changed among the latter group during that time period (Fagan and Huang 2019).

Responses by industry, business groups, and governmental agencies at all levels was, and continues to be, very complex. They have vacillated from strong disagreement (denial) and serious doubts (skepticism) about climate change and its causes to more or less tentative acceptance and enthusiastic support for action. A more detailed analysis of these responses is provided in Chapter 2 of this book.

Conclusion

Humans have been interested in the general features of Earth's climate for millennia. For most of that time, research has focused on specific elements of the climate, such as changes in temperature and the appearance and disappearance of massive glaciation (ice ages). Climatology as a science began to develop only in the mid-20th century, with the development of methods to obtain more precise data on the abundance of carbon dioxide in the atmosphere and the oceans, the influence of other greenhouse gases on climate, the effect of human activities on climate, and related topics. In today's world, the reality of climate change and anthropogenic factors affecting the change are understood and accepted by a large fraction of the community of professional climate researchers. There really is no longer any "debate" among researchers in the field as to whether climate is changing to a significant degree and what the role of human activities play in that phenomenon. A similar consensus among the general public and public officials does not yet exist. In fact, many issues about climate change remain to be resolved before actions for dealing with the phenomenon can be identified and implemented. The direction of this debate is the topic of Chapter 2 of this book.

References

"About Ice Cores." n.d. Ice Core Facility. National Science Foundation. https://icecores.org/about-ice-cores. Accessed on August 18, 2019.

"The Andean-Saharan Ice Age, Ep 3 History of Climate Change CO_2 and Human Evolution." 2018. YouTube. https://www.youtube.com/watch?v=1rPkPKPyGXY. Accessed on August 8, 2019.

Arrhenius, Svante. 1896. "On the Influence of Carbonic Acid in the Air upon the Temperature of the Ground." *The London, Edinburgh, and Dublin Philosophical*

Magazine and Journal of Science 41(251): 237–276. doi:10.1080/14786449608620846. http://empslocal.ex .ac.uk/people/staff/gv219/classics.d/Arrhenius96.pdf. Accessed on August 12, 2019.

Barry, Roger G. 2013. "A Brief History of the Terms Climate and Climatology." *International Journal of Climatology* 33: 1317–1320. https://rmets.onlinelibrary.wiley.com/doi/pdf /10.1002/joc.3504. Accessed on August 27, 2019.

Bekker, Andrey. 2014. "Huronian Glaciation." *Encyclopedia of Astrobiology*. Berlin: Springer-Verlag. https://link.springer .com/content/pdf/10.1007/978-3-642-27833-4_742-4 .pdf. Accessed on August 6, 2019.

Benn, Hilary. 2007. "'The Most Important Problem We Face.'" *The Guardian*. https://www.theguardian.com /commentisfree/2007/may/14/howseriousaproblemis. Accessed on August 5, 2019.

Berger, A., et al. 2016. "Interglacials of the Last 800,000 Years." *Reviews of Geophysics* 54(1): 162–219. https:// agupubs.onlinelibrary.wiley.com/doi/pdf/10.1002 /2015RG000482. Accessed on August 7, 2019.

Berger, André. 1984. *Milankovitch and Climate: Understanding the Response to Astronomical Forcing*. Dordrecht, Netherlands: D. Reidel.

Bhatia, Rehemat. 2013. "Diatoms and Climate." Diatoms Online. http://diatoms.myspecies.info/node/202. Accessed on August 20, 2019.

Black, Richard. 2011. "Tyndall's Climate Message, 150 Years On." BBC News. https://www.bbc.com/news/science -environment-15093234. Accessed on August 11, 2019.

Blom, Philipp. 2019. *Nature's Mutiny: How the Little Ice Age Transformed the West and Shaped the Present*. London: Picador.

Broecker, W. S. 2002. *The Glacial World According to Wally*. 3rd rev. ed. Palisades, NY: Lamont-Doherty Earth

Observatory of Columbia University. https://www.ldeo
.columbia.edu/~broecker/Home_files/GlacialWorld.pdf.
Accessed on August 25, 2019.

Budyko, M. I. 1969. "The Effect of Solar Radiation Variations
on the Climate of the Earth." *Tellus* 21(5): 611–619.
https://onlinelibrary.wiley.com/doi/pdf/10.1111/j.2153
-3490.1969.tb00466.x. Accessed on August 25, 2019.

Cable, Richard. 2009. "'Since Records Began': A Brief Guide
to Who's Taking the Temperature." BBC. https://www.bbc
.co.uk/blogs/climatechange/2009/03/since_records_began
_a_brief_gu.html. Accessed on August 5, 2019.

Callendar, G. S. 1938. "The Artificial Production of Carbon
Dioxide and Its Influence on Temperature." *Quarterly
Journal of the Royal Meteorological Society* 64(275): 223–
240. http://www.met.reading.ac.uk/~ed/callendar_1938
.pdf. Accessed on August 16, 2019.

"Causes of Climate Change." 2019. Climate Science
Investigations. NASA. http://www.ces.fau.edu/nasa/module
-4/causes-2.php. Accessed on August 9, 2019.

Chamberlin, Thomas Chrowder. 1908. "Soil Wastage." The
Popular Science Monthly. https://en.wikisource.org/wiki
/Popular_Science_Monthly/Volume_73/July_1908/Soil
_Wastage. Accessed on August 16, 2019.

"Chemical Proxies Finder." 2016. http://climateproxiesfinder
.ipsl.fr/. Accessed on August 5, 2019.

"Climate." 2017. Barvenon.com. http://www.barvennon.com
/climate.html. Accessed on August 8, 2019.

"Climate Models." n.d. Climate.gov. NOAA. https://www
.climate.gov/maps-data/primer/climate-models. Accessed
on August 21, 2019.

Climate Research Committee. National Research Council
Staff. 1996. *Natural Climate Variability on Decade-to-
Century Time Scales*. Washington, DC: National Academies
Press.

Cook, John. 2019. "What Ended the Little Ice Age?" Skeptical Scientist. https://skepticalscience.com/coming -out-of-little-ice-age.htm. Accessed on August 8, 2019.

"CO_2 at NOAA's Mauna Loa Observatory Reaches New Milestone: Tops 400 ppm." 2013. Global Monitoring Division. Earth System Research Laboratory. https://www .esrl.noaa.gov/gmd/news/7074.html. Accessed on August 11, 2019.

Crimmins, Mike. n.d. "An Introduction to Global Climate Modeling." University of Arizona. https:// d3dqsm2futmewz.cloudfront.net/docs/dcdc/website /documents/MikeCrimmins_DCDC2013.pdf. Accessed on August 26, 2019.

DeConto, Robert M. 2009. "Plate Tectonics and Climate Change." In *Encyclopedia of Paleoclimatology and Ancient Environments*, edited by Vivien Gornitz, 784–798. Dordrecht, Netherlands; New York: Springer. https:// www.geo.umass.edu/climate/papers2/deconto_tectonics &climate.pdf. Accessed on August 9, 2019.

de Solla Price, Derek J. [1963] 1986. *Little Science, Big Science*. New York: Columbia University Press. http://www .andreasaltelli.eu/file/repository/Little_science_big_science _and_beyond.pdf. Accessed on August 21, 2019.

"Developers of Climate Models." 2012. Climate Modeling 101. https://nas-sites.org/climate-change/climatemodeling /page_6_1.php. Accessed on August 26, 2019.

Dewey, Martina M. 2010. "Paleo Proxy Data: What Is It?" IEDRA. http://iedro.org/articles/paleo-proxy-data/. Accessed on August 5, 2019.

Dittmann, Roger. 1987. "Book Review. Beyond the Laboratory: Scientists as Political Activists in 1930s America. Peter J. Kuznick, University of Chicago Press (1987)." https://physics.fullerton.edu/~rdittmann/USFSS

/Beyond%20the%20Laboratory.htm. Accessed on August 21, 2019.

Douglas, Sir William Sholto. 2018. "RADAR: The Battle Winner?" Royal Air Force Museum. https://www .rafmuseum.org.uk/research/online-exhibitions/history -of-the-battle-of-britain/radar-the-battle-winner.aspx. Accessed on August 21, 2019.

Emiliani, Cesare. 1958. "Ancient Temperatures." *Scientific American* 198(2): 54–66.

Emiliani, Cesare. 1966. "Isotopic Paleotemperatures." *Science* 154(3751): 851–857.

Emiliani, Cesare. 1972. "Quaternary Hypsithermals." *Quaternary Research* 2(3): 270–273.

Evans, Jon. 2013. "Chemical Climate Proxies." Chemistry World. https://www.chemistryworld.com/features/chemical -climate-proxies/5817.article. Accessed on August 5, 2019.

Fagan, Brian. 2019. *Little Ice Age: How Climate Made History 1300–1850.* New York: Basic Books.

Fagan, Moira, and Christine Huang. 2019. "A Look at How People around the World View Climate Change." Pew Research Center. https://www.pewresearch.org/fact-tank /2019/04/18/a-look-at-how-people-around-the-world-view -climate-change/. Accessed on August 27, 2019.

Goosse, H., et al. 2008–2010. "Modeling the Climate System." Introduction to Climate Dynamics and Climate Modeling. http://www.climate.be/textbook. Accessed on August 21, 2019.

Gornitz, Vivien. 2015. "Paleoclimate Proxies, An Introduction." In *Encyclopedia of Paleoclimatology and Ancient Environments*, edited by Vivien Gornitz, 716–721. Dordrecht, Netherlands: Springer.

Hegerl, Gabriele C., et al. 2007. "Understanding and Attributing Climate Change." In *Climate Change 2007:*

The Physical Science Basis, edited by Susan Solomon et al., 663–745. Cambridge, UK: Cambridge University Press.

Hessler, Angela M. 2011. "Earth's Earliest Climate." *Nature Education Knowledge* 3(10): 24. https://www.nature .com/scitable/knowledge/library/earth-s-earliest-climate -24206248. Accessed on August 5, 2019.

Hickman, Leo. 2018. "Timeline: The History of Climate Modeling." CarbonBrief. https://www.carbonbrief.org /timeline-history-climate-modelling. Accessed on August 26, 2019.

Hirst, K. Kris. 2019. "Palynology Is the Scientific Study of Pollen and Spores." Thought Co. https://www.thoughtco .com/palynology-archaeological-study-of-pollen-172154. Accessed on August 19, 2019.

"A History of Climate Activities." 2009. World Meteorological Organization. https://public.wmo.int/en /bulletin/history-climate-activities. Accessed on August 26, 2019.

"History of the National Weather Service." n.d. National Weather Service. https://www.weather.gov/timeline. Accessed on August 5, 2019.

"How Does Carbon Dating Work?" 2019. Beta Analytic. https://www.radiocarbon.com/about-carbon-dating.htm. Accessed on August 18, 2019.

"How to Talk to an Ostrich: 'Who Says CO_2 Heats Things Up?'" 2012. Earth: The Operator's Manual. https://www .youtube.com/watch?v=JoR4ezwKh5E&feature=autoplay &list=UUShAg7p1e30Bk-XWS7EiCgQ&playnext=1. Accessed on August 21, 2019 (video).

Huang, Shaopeng, Henry N. Pollack, and Po-Yu Shen. 2000. "Temperature Trends over the Past Five Centuries Reconstructed from Borehole Temperatures." *Nature* 403(6771): 756–758. https://deepblue.lib.umich.edu

/bitstream/handle/2027.42/62610/403756a0.pdf. Accessed on August 18, 2019.

"Implications of Rising Carbon Dioxide Content of the Atmosphere." 1963. The Conservation Foundation. https:// babel.hathitrust.org/cgi/pt?id=mdp.39015004619030 &view=1up&seq=3. Accessed on August 26, 2019.

"Is an Ice Age Coming?" 2016. PBS Space Time. https://www .youtube.com/watch?time_continue=8&v=ztninkgZ0ws. Accessed on August 9, 2019 (video).

Kaplan, Jed O., Kristen M. Krumhardt, and Niklaus Zimmermann. 2009. "The Prehistoric and Preindustrial Deforestation of Europe." *Quaternary Science Reviews* 28: 3016–3034. https://www.wsl.ch/staff/niklaus.zimmermann /papers/QuatSciRev_Kaplan_2009.pdf. Accessed on August 11, 2019.

Keeling, Charles D. 1960. "The Concentration and Isotopic Abundances of Carbon Dioxide in the Atmosphere." *Tellus* 12(2): 200–203. https://onlinelibrary.wiley.com/doi/epdf /10.1111/j.2153-3490.1960.tb01300.x. Accessed on August 22, 2019.

Kershaw, Tristan. 2017. "Climate Change and Its Impacts." IOP Publishing. https://iopscience.iop.org/chapter/978-0 -7503-1197-7/bk978-0-7503-1197-7ch1.pdf. Accessed on August 9, 2019.

Kopp, Robert E., et al. 2005. "The Paleoproterozoic Snowball Earth: A Climate Disaster Triggered by the Evolution of Oxygenic Photosynthesis." *Proceedings of the National Academy of Sciences of the United States of America* 102(32): 11131–11136. https://www.pnas.org/content/pnas/102/32 /11131.full.pdf. Accessed on August 6, 2019.

Landsberg, Helmut E. 1970. "Man-Made Climatic Changes." *Science* 170(3964): 1265–1274.

Lemmons, Richard. 2019. "Climate Change and Subsurface Temperature." Climate Policy Watcher. https://www

.climate-policy-watcher.org/temperature-logs/climate
-change-and-subsurface-temperature.html. Accessed on
August 18, 2019.

Mann, Michael E. 2015. *The Hockey Stick and the Climate
Wars: Dispatches from the Front Lines*. New York: Columbia
University Press.

Marks, Kelley. 2019. "21 Ways Archeologists Date Ancient
Artifacts." HubPages. https://hubpages.com/education
/How-Do-Archaeologists-Date-Past-Events. Accessed on
August 17, 2019.

Martinez, Lori. [1996] 2000. "Crossdating: The Basic
Principle of Dendrochronology." University of Arizona.
https://www.ltrr.arizona.edu/lorim/basic.html. Accessed on
August 19, 2019.

Mason, John. 2012. "Two Centuries of Climate Science:
Part One—Fourier to Arrhenius, 1820–1930." Skeptical
Scientist. https://skepticalscience.com/two-centuries
-climate-science-1.html. Accessed on August 16, 2019.

Mason, Matthew. 2019. "With Palynology We Can See the
Tiniest Details." Environmental Science. https://www
.environmentalscience.org/palynology. Accessed on August
19, 2019.

McKitrick, Ross R. n.d. "Global Warming: Paleoclimate /
Hockey Stick." https://www.rossmckitrick.com/paleo
climatehockey-stick.html. Accessed on August 23, 2019.

McLamb, Eric. 2011. "Earth's Beginnings: The Origins of
Life." Ecology. http://www.ecology.com/2011/09/10/earths
-beginnings-origins-life/. Accessed on August 5, 2019.

McNeill, Lelia. 2016. "This Lady Scientist Defined the
Greenhouse Effect But Didn't Get the Credit, Because
Sexism." Smithsonian.com. https://www.smithsonianmag
.com/science-nature/lady-scientist-helped-revolutionize
-climate-science-didnt-get-credit-180961291/#MCT
gb0VrWmAcDkIx.99. Accessed on August 11, 2019.

"Milankovitch Cycles III: Precession of the Seasons, Seasonality, and Extent of Glaciation." 2006. Some Fundamentals of Mineralogy and Geochemistry. http:// railsback.org/Fundamentals/SFMGMilankovitch3 -Precession01.pdf. Accessed on August 9, 2019.

Montanez, Isabel P., and Christopher J. Poulsen. 2013. "The Late Paleozoic Ice Age: An Evolving Paradigm." *Annual Review of Earth and Planetary Sciences* 41(1): 629–656. https://pdfs.semanticscholar.org/dd43/d77e91b131b2de3b 5fa4ffb168ca429e7c94.pdf. Accessed on August 7, 2019.

Muller, Richard. 2004. "Global Warming Bombshell." MIT Technology Review. https://www.technologyreview.com/s /403256/global-warming-bombshell/. Accessed on August 23, 2019. Accessed on August 23, 2019.

Nave, Rod. 2017. "Clocks in the Rocks." Hyperphysics. http://hyperphysics.phy-astr.gsu.edu/hbase/Nuclear/clkroc .html. Accessed on August 18, 2019.

Novus RCN. 2014. "Dendrochronology: How to Core a Tree." YouTube. https://www.youtube.com/watch?v= jPJUewNcvao. Accessed on August 19, 2019.

Ocko, Illissa. 2019. "9 Ways We Know Humans Triggered Climate Change." Environmental Defense Fund. https:// www.edf.org/climate/9-ways-we-know-humans-triggered -climate-change. Accessed on August 5, 2019.

Ogurtsov, Maxim, et al. 2015. *The Sun-Climate Connection over the Last Millennium Facts and Questions.* Sharjah, UAE: Bentham Science Publishers.

Oreskes, Naomi. 2004. "Beyond the Ivory Tower: The Scientific Consensus on Climate Change." *Science* 306(5702): 1686. https://science.sciencemag.org/content /306/5702/1686. Accessed on August 27, 2019. For an expanded discussion of this topic, see Naomi Oreskes. 2007. "The Scientific Consensus on Climate Change: How Do We Know We're Not Wrong?" In *Climate Change: What*

It Means for Us, Our Children, and Our Grandchildren,
edited by Joseph F. C. DiMento and Pamela Doughman,
65–99. Cambridge, MA: MIT Press. http://citeseerx.ist.psu
.edu/viewdoc/download?doi=10.1.1.800.4969&rep=rep1
&type=pdf. Accessed on December 6, 2019.

"Paleoclimatology Datasets." n.d. National Centers for
Environmental Information. https://www.ncdc.noaa.gov
/data-access/paleoclimatology-data/datasets. Accessed on
August 20, 2019.

"Palynology." n.d. PopFlock. http://www.popflock.com/learn
?s=Palynology. Accessed on August 19, 2019.

Pearson, Paul. 2012. "Oxygen Isotopes in Foraminifera:
Overview and Historical Review." *The Paleontological
Society Papers* 18: 1–38. https://orca.cf.ac.uk/41988/1
/Pearson%202012%20Oxygen%20isotope%20review.pdf.
Accessed on August 6, 2019.

Pidcock, Roz. 2015. "The Most Influential Climate Change
Papers of All Time." CarbonBrief. https://www.carbonbrief
.org/the-most-influential-climate-change-papers-of-all
-time. Accessed on August 26, 2019.

Plass, Gilbert N. 1956. "The Carbon Dioxide Theory of
Climatic Change." *Tellus* 8: 149–154. https://onlinelibrary
.wiley.com/doi/epdf/10.1111/j.2153-3490.1956.tb01206
.x. Accessed on August 21, 2019.

Pospelova, Vera, Thomas F. Pedersen, and Anne de Vernal.
2006. "Dinoflagellate Cysts as Indicators of Climatic
and Oceanographic Changes During the past 40 kyr
in the Santa Barbara Basin, Southern California."
Paleoceanography 21(2). doi:10.1029/2005PA001251.
https://agupubs.onlinelibrary.wiley.com/doi/epdf/10.1029
/2005PA001251. Accessed on August 20, 2019.

Poushter, Jacob, and Christine Huang. 2019. "Climate
Change Still Seen as the Top Global Threat, but
Cyberattacks a Rising Concern." Pew Research Center.

https://www.pewresearch.org/global/2019/02/10/climate
-change-still-seen-as-the-top-global-threat-but-cyberattacks
-a-rising-concern/. Accessed on August 27, 2019.

"Preparing Reports." 2019. Intergovernmental Panel
on Climate Change. https://www.ipcc.ch/about
/preparingreports/. Accessed on August 26, 2019.

Railsback, Bruce, et al. 2015. "An Optimized Scheme of
Lettered Marine Isotope Substages for the Last 1.0 Million
Years, and the Climatostratigraphic Nature of Isotope
Stages and Substages." *Quaternary Science Reviews* 111:
94–106.

"Report of the International Conference on the Assessment
of the Role of Carbon Dioxide and of Other Greenhouse
Gases in Climate Variations and Associated Impacts."
1986. UMO- No. 661. https://library.wmo.int/doc_num
.php?explnum_id=8512. Accessed on August 26, 2019.

"Reports." 2019. Intergovernmental Panel on Climate
Change. http://www.ipcc.ch/reports/. Accessed on August
26, 2019.

Revelle, Roger, and Hans E. Suess. 1957. "Carbon Dioxide
Exchange between Atmosphere and Ocean and the
Question of an Increase of Atmospheric CO_2 during
the Past Decades." *Tellus* 9(1): 18–27. https://pdfs
.semanticscholar.org/d014/06a57bff758203390e36247bd9
6e0c9f8102.pdf. Accessed on August 21, 2019.

Richard, Kenneth. 2017. "Changing Scientific Consensus."
CO_2 Coalition. https://co2coalition.org/2017/11/28/before
-1960s-70s-global-cooling-was-erased-it-caused-droughts
-crop-failures-glacier-advance-ice-age-threats/. Accessed on
August 25, 2019.

Rosby, C.-G. 1959. "Current Problems in Meteorology."
In *The Atmosphere and the Sea in Motion*, edited by Berg
Bolin, 9–50. New York: Rockefeller Institute Press. https://
math.nyu.edu/~gerber/courses/2018-fruhling/bolin_etal

-atmoshere_sea_in_motion-1959.pdf. Accessed on August 21, 2019.

Sample, Ian. 2005. "The Father of Climate Change." *The Guardian.* https://www.theguardian.com/environment /2005/jun/30/climatechange.climatechangeenvironment2. Accessed on August 16, 2019.

Schriber, Michael. 2015. "'Snowball Earth' Might Be Slushy." Astrobiology Magazine. https://www.astrobio.net/news -exclusive/snowball-earth-might-be-slushy/. Accessed on August 6, 2019.

Sellers, W. D. 1965. *Physical Climatology.* Chicago: University of Chicago Press.

"Sense Seeker." 2012. Group Dynamics. *The Economist.* https://www.economist.com/node/21556212/comments ?page=4. Accessed on August 5, 2019.

"1700 Years of Global Temperature Change from Proxy Data." 2017. GlobalChange.gov. https://data.globalchange .gov/report/climate-science-special-report/chapter/our -changing-climate/figure/temp-change-from-proxy-data. Accessed on August 23, 2019. (Enlarged graph available at bottom of page.)

Sheppard, Paul R. 2014. "Crossdating Tree Rings." University of Arizona. https://www.ltrr.arizona.edu/skeletonplot /introcrossdate.htm. Accessed on August 19, 2019.

Shindell, Drew T., et al. 2001. "Solar Forcing of Regional Climate Change during the Maunder Minimum." *Science* 294(5549): 2149–2152. http://www.meteo.psu.edu /holocene/public_html/shared/articles/Shindelletal01.pdf. Accessed on August 10, 2019.

Snyder, Carolyn W. 2016. "Evolution of Global Temperature over the Past Two Million Years." *Nature* 538(7624): 226–228.

Soon, Willie, and Sallie Baliunas. 2003. "Proxy Climatic and Environmental Changes of the Past 1000 Years." *Climate*

Research 23: 89–110. https://www.int-res.com/articles
/cr2003/23/c023p089.pdf. Accessed on August 25, 2019.

Starr, Victor P., and Abraham H. Oort. 1973. "Five-Year
Climatic Trend for the Northern Hemisphere." *Nature*
242(5396): 310–313.

Stocker, T. 2016. "Introduction to Climate Modeling."
Physikalisches Institut. Universität Bern. https://
climatehomes.unibe.ch/~stocker/papers/stocker16icm.pdf.
Accessed on August 26, 2019.

"Tree-Ring Dating." 2013. Time Team America. http://
www.pbs.org/time-team/experience-archaeology
/dendrochronology/. Accessed on August 19, 2019.

Trump, Donald. 2012. Twitter. https://twitter.com
/realdonaldtrump/status/265895292191248385?lang=en.
Accessed on August 5, 2019.

Tyndall, John. 1869. *Heat Considered as a Mode of
Motion.* New York: D. Appleton. https://ia800704
.us.archive.org/28/items/heatconsideredas00tyndrich
/heatconsideredas00tyndrich_bw.pdf. Accessed on August
12, 2019.

Ueda, Herbert T., and Pavel G. Talalay. 2007. "Fifty Years
of Soviet and Russian Drilling Activity in Polar and
Non-Polar Ice: A Chronological History." Cold Regions
Research and Engineering Laboratory. U.S. Army Engineer
Research and Development Center. https://apps.dtic.mil
/dtic/tr/fulltext/u2/a472548.pdf. Accessed on August 25,
2019.

"UNFCCC Process-and-Meetings." 2019. United Nations
Climate Change. https://unfccc.int/process-and-meetings
#:2cf7f3b8-5c04-4d8a-95e2-f91ee4e4e85d. Accessed on
August 27, 2019.

"United Nations Framework Convention on Climate
Change." 1992. United Nations. https://unfccc.int/resource
/docs/convkp/conveng.pdf. Accessed on August 26, 2019.

Verheggen, Bart. 2016. "Earth's Temperature over the Past Two Million Years." My View on Climate Change. https://ourchangingclimate.wordpress.com/2016/10/06/earths-temperature-over-the-past-two-million-years-snyder-earth-system-sensitivity/. Accessed on August 7, 2019.

Watts, Anthony. 2013a. "Dr. Vincent Gray on Historical Carbon Dioxide Levels." Watts Up with That. https://wattsupwiththat.com/2013/06/04/dr-vincent-gray-on-historical-carbon-dioxide-levels/. Accessed on August 6, 2019.

Watts, Anthony. 2013b. "The 1970's Global Cooling Compilation—Looks Much Like Today." Watts Up with That? https://wattsupwiththat.com/2013/03/01/global-cooling-compilation/. Accessed on August 25, 2019.

Weart, Spencer. 2007. "Roger Revelle's Discovery." The Discovery of Global Warming. American Institute of Physics. https://history.aip.org/climate/Revelle.htm. Accessed on August 21, 2019.

Weart, Spencer. 2017. "The Public and Climate Change." The Discovery of Global Warming. American Institute of Physics. https://history.aip.org/climate/public.htm. Accessed on August 21, 2019.

Weart, Spencer. 2018. "Simple Models of Climate Change." The Discovery of Global Warming. American Institute of Physics. https://history.aip.org/climate/simple.htm#L_M006. Accessed on August 16, 2019.

Weart, Spencer. 2019a. "Climatology as a Profession." The Discovery of Global Warming. American Institute of Physics. https://history.aip.org/climate/climogy.htm. Accessed on August 27, 2019.

Weart, Spencer. 2019b. "General Circulation Models of Climate." The Discovery of Global Warming. American Institute of Physics. https://history.aip.org/climate/GCM.htm. Accessed on August 21, 2019.

Wetmore, Karen. n.d. "Foram Facts—An Introduction to Foraminifera." https://ucmp.berkeley.edu/fosrec/Wetmore .html. Accessed on August 20, 2019.

"What's the Difference between Weather and Climate? Climate Is What You Expect, Weather Is What You Get." 2018. National Centers for Environmental Information. https://www.ncei.noaa.gov/news/weather-vs-climate. Accessed on August 5, 2019.

"World Climate Programme." 2019. World Meteorological Organization. http://www.wmo.int/pages/prog/wcp/index _en.html. Accessed on August 26, 2019.

Yarbrough, Scott. 2015. *The New Ice Age: The Truth about Climate Change*. n.p.: Scott Yarbrough.

Zhao, Jing, et al. 2014. "Late Cretaceous Palynology and Paleoclimate Change: Evidence from the SK1 (South) Core, Songliao Basin, NE China." *Science China Earth Sciences* 57(12): 2985–2997.

Scientists are normally not very political animals. They generally prefer to stay in their own laboratories, where they focus on their own research interests. They share and discuss their findings almost entirely with other workers in the same field. They tend not to write articles for the general public in the popular media. Nor do they exert pressure on politicians to take some form of action on the discoveries they make. Some small confirmation of this tendency comes from the fact that, as of early 2020, only four members of the U.S. Congress could be classified as research scientists. Another four dozen came from science-related fields, such as nursing, medicine, veterinary medicine, engineering, and computer science ("US 116th Congress Sets New Record for Members with STEM Backgrounds" 2019). Under the most generous definition of "scientist," then, about 9 percent of senators and representatives can be given that label.

Climate Change as a National Issue

Exceptions to that rule exist, however. In a handful of cases in history, individuals or groups of scientists have become so concerned about the direction of their research that they decided

A wildfire starts by June Lake Loop in California's Eastern Sierras. Climate change is expected to increase the intensity of such fires in the near future. (Monica Flynn/Dreamstime.com)

it was critical to alert the public and policy makers. Such was the case, for example, at the end of World War II when many scientists who had worked on the development of the atomic bomb realized the issues their research was likely to have in the postwar world. They organized to educate and lobby politicians about the potential risks and benefits of nuclear power, actions that eventually led to the way in which that field is operated and monitored even today. (A very rich literature exists about the interplay of pure science and politics around the time of World War II. One of the best books available is Greenberg [1969] 1999.)

Climate science is another field for which an exception to the general rule exists. As early as 1965, a group of climate scientists expressed their concerns about global warming in a report commissioned by President Lyndon Johnson, "Restoring the Quality of Our Environment." In a lengthy appendix to that report, five eminent climate researchers—Roger Revelle, Wallace Broecker, Harmon Craig, C. D. Keeling, and J. Smagorinsky—laid out the evidence for global warming, reviewed likely anthropogenic causes for the phenomenon, and discussed several possible effects from increased carbon-dioxide levels in the atmosphere. In its conclusion to the appendix, the authors penned one of the most famous comments on the climate change problem at the time: "Man is unwittingly conducting a vast geophysical experiment" ("Restoring the Quality of Our Environment. Report of The Environmental Pollution Panel. President's Science Advisory Committee" 1965).

Every president since the release of this report has been made aware of a potential climate change crisis, with varying effect in each case. In 1969, for example, an aide to President Richard Nixon sent a memo to the president warning about possible consequences of global warming. In his memo, Daniel Moynihan noted that "[i]t is now pretty clearly agreed that the CO_2 content will rise 25% by 2000. This could increase the average temperature near the earth's surface by 7 degrees Fahrenheit. This in turn could raise the level of the sea by 10 feet. Goodbye

New York. Goodbye Washington, for that matter" ("For John Ehrlichman" 1969).

The 1970s were a decade of rapidly increasing interest in environmental issues. That interest was reflected on a national level by the creation of many new governmental organizations designed to deal with one or more aspects (and, in a few cases, all) of the nation's environmental problems. A number of those agencies lasted only a few years or evolved in other newer agencies, most of which have survived in one form or another to the present day. Among the agencies that were assigned some aspect of the climate change issue were the National Oceanic and Atmospheric Administration; Environmental Protection Agency; National Aeronautics and Space Administration; and Department of Energy. (A superb review of the role of climate policy in the United States after the mid-1950s can be found at Weart 2019; also see Lewin 2010.)

The period also saw the first efforts to develop a dedicated federal program for climate change research. The result of that effort was the passage in 1978 of the National Climate Program Act. The purpose of that act was to provide funding for research on and dissemination of information about climate change. The enthusiasm that led to adoption of the act was not matched by efforts to achieve its goal, and it is generally regarded largely as a failure. It was, nonetheless, the first major effort by the U.S. Congress to deal with global warming and climate change issues (Epstein 1978).

By the 1980s, a handful of senators and representatives had become committed to developing methods by which the federal government could promote climate change research and education (for more on this topic, see Rich 2018). One of the most important features of these efforts was the opportunity it gave for climate researchers to appear before congressional committees, report on their activities, and repeat warnings about the risks of ignoring climate change. These appearances were often reported in the mass media, where the general public could have a taste of developments in climate science. One example is the appearance in June 1988 of climatologist James Hansen before the U.S. Senate Committee on Energy and Natural Resources. A frequent guest of congressional committees on the topic of climate change, Hansen introduced his testimony with the observation that

> I would like to draw three main conclusions. Number one, the earth is warmer in 1988 than at any time in the history of instrumental measurements. Number two, the global warming is now large enough that we can ascribe with a high degree of confidence a cause and effect relationship to the greenhouse effect. And number three, our computer climate simulations indicate that the greenhouse effect is already large enough to begin to effect the probability of extreme events such as summer heat waves. . . . [In addition:] the warming is larger than natural climate variability. ("Statement of Dr. James Hansen, Director, NASA Goddard Institute for Space Sciences" 1988)

In an interview with Hansen shortly after his testimony, the *New York Times* reported to the general public that experts were now "99 percent certain that the [global] warming trend was not a natural variation but was caused by a buildup of carbon dioxide and other artificial gases in the atmosphere." The country had been warned that a major disaster was possible in the not-too-distant future (Shabecoff 1988).

Over the past three decades, a pattern has evolved in the discussion over global warming and climate change. Experts in the field have become increasingly certain about global warming and its possible consequences, as well as its result largely of human activities. Members of the general public, especially including politicians and other policy makers, have either accepted or rejected that conclusion, generally speaking, along political party lines. That is, throughout this period, Democrats and those who lean Democratic have generally agreed with experts, while Republicans and those who lean Republican tend to doubt or deny these conclusions. (This statement is not a commentary on the virtues or defects of one public view or another. It simply reflects the results of public opinion polling throughout this period; Kennedy and Hefferon 2019.)

The actions of the two most recent presidential administrations—that of Barack Obama and that of Donald Trump—tend to confirm this trend. On the one hand, Obama's term in office was marked by a variety of activities designed to more completely involve the government in the process of reducing climate change. Among the actions taken include the following:

2013: President Obama announces his Climate Action Plan, consisting of a variety of activities, such as promoting the development of renewable energy; reducing pollution from fossil fuel energy plants; increasing fuel economy standards; cutting energy waste in homes, businesses, and factories; reducing greenhouse gases other than carbon dioxide, such as methane and chlorofluorohydrocarbons; and developing other federal programs for an increased use of clean energy in the nation ("The President's Climate Action Plan" 2013).

2013: The Better Buildings Challenge is created to encourage energy-efficient upgrades in commercial structures.

2014: A presidential memo requires all federal agencies to conduct a Quadrennial Energy Review for the development of a comprehensive strategy for the infrastructure involved in transporting, transmitting, and delivering energy ("Obama Administration Launches Quadrennial Energy Review" 2014).

2014: The Climate Data Initiative is created to encourage local communities to develop structures and systems that can be used to deal with future climate change problems

("The President's Climate Data Initiative: Empowering America's Communities to Prepare for the Effects of Climate Change" 2014).

2015: The Environmental Protection Agency and Obama administration announce a new Clean Power Plan. The goal of the plan is to increase the emphasis on the use of renewable energy by the nation's industrial sector, accompanied by a significant reduction in its dependence on fossil fuels as a source of energy ("Overview of the Clean Power Plan" 2015).

2016: On September 3, President Obama signs documents that formally enroll the United States in the Paris Climate Agreement. He says that "someday we may see this as the moment that we finally decided to save our planet" (Somanander 2016; for details on the Paris Agreement, see "The Paris Agreement" 2019).

This is a partial list of actions taken between 2009 and 2016 by the Obama administration (for more detail, see Gutin and Ingargiola 2015).

When President Donald Trump took office in January 2016, he arrived with a different agenda for the climate change problem. Among the actions he took early in his administration were the following (as of this writing, President Trump has been in office for a much shorter term than President Obama, so that the following list is not comparable with the Obama list):

2017: In a Presidential Executive Order issued two months after his inauguration, Trump rescinds about 20 of Obama's Executive Orders, memos, and other actions dealing with climate change ("Presidential Executive Order on Promoting Energy Independence and Economic Growth" 2017).

2017: Trump announces that the United States will withdraw from the Paris Climate Agreement. Under conditions of the treaty, the action will take effect on November 4, 2020, one day after the presidential election of that year ("Statement by President Trump on the Paris Climate Accord" 2017).

2018: Obama's rigorous fuel efficiency rule of 2012 is revoked and replaced by a new rule, the Safer Affordable Fuel-Efficient (SAFE) Vehicles Rule for Model Years 2021–2026, relaxing fuel efficiency standards ("U.S. DOT and EPA Propose Fuel Economy Standards for MY 2021–2026 Vehicles" 2018).

2019: The Trump administration revokes Obama's Clean Power Plan, issued in 2015, and replaces it with a more fossil-fuel-energy friendly Affordable Clean Energy rule ("EPA Finalizes Affordable Clean Energy Rule, Ensuring Reliable, Diversified Energy Resources while Protecting our Environment" 2019).

2019: An Obama rule requiring oil and gas companies to reduce methane emissions is revoked and replaced by a much less rigorous requirement for the escape of methane from such operations ("Oil and Natural Gas Sector: Emission Standards for New, Reconstructed, and Modified Sources Review" 2019).

The primary lesson to be learned from this review is that climate change is far more than a scientific issue. No matter what experts in the field of climatology may believe, actual responses by national, state, and local governments depends on a host of other factors, probably most important among them being political forces. That lesson is critical to all of the specific climate change issues discussed later in this chapter.

Possible Consequences of Climate Change

At the heart of the climate change debate in the 21st century is the question, "What difference will/might climate change for the natural environment and human civilization?" As information about global warming and climate change has increased in quantity and sophistication, climatologists have moved beyond the question as to *whether or not climate change is occurring* to *what effects are these changes likely to produce on the planet?*

A very substantial amount of research has been devoted to this question. Rather than wading through the many thousands of studies on various aspects of this question, the interested observer can turn to a handful of major reports that summarize and comment on these studies. The most important of those reports are the five assessment reports produced by the IPCC over the last 30 years. Each report runs many hundreds of pages, but some general conclusions can be noted here (the IPCC reports are available in full at "Reports" 2019).

One essential feature of the IPCC approach is its so-called "likelihood scale." The scale reflects the confidence with which each conclusion has been drawn. For example, scientists can measure the temperature of Earth's atmosphere with a high degree of accuracy today. Records of those temperatures over the past hundred years are highly reliable. We really do have a true record of atmospheric temperatures over the time. Of course, there will always be individuals who are suspicious of any "fact" scientists put forward. There are still people who believe the Earth is flat or that the 9/11 terrorist attack never occurred. But the vast majority of people are willing to accept as true many of the observations that scientists have recorded about the Earth.

Other information falls into a category that might be called *almost certainly true*, or *virtually certain*. After all, there are many characteristics of Earth that scientists simply can't measure directly. For example, what was Earth's annual average temperature 1,000 years ago, 10,000 years ago, or 100,000 years ago? There is no way to answer those questions without a doubt. But researchers have now developed methods for estimating these temperatures with a high degree of accuracy. For example, the chemical properties of ice differ to some extent depending on the temperature at which it is formed. So, studying ice cores

Table 2.1   Likelihood Scale (IPCC)

Term	Likelihood of the Outcome
Virtually certain	99–100% probability
Very likely	90–100% probability
Likely 66–100% probability	About as likely as not 33 to 66% probability
Unlikely 0–33% probability	Very unlikely 0–10% probability
Exceptionally unlikely	0–1% probability

Source: Mastrandrea, Michael D., et al. 2010. "Guidance Note for Lead Authors of the IPCC Fifth Assessment Report on Consistent Treatment of Uncertainties," Table 1. IPCC Cross-Working Group Meeting on Consistent Treatment of Uncertainties. https://wg1.ipcc.ch/SR/documents/ar5_uncertainty-guidance-note.pdf. Accessed on May 29, 2019. Used by permission of the IPCC.

that date back thousands of years provides an indirect measure of Earth's temperature at the time.

The IPCC has developed a formal system for classifying the statements it makes about possible future consequences of climate change. As shown in table 2.1, that system consists of five categories, ranging from "virtually certain" to "extremely unlikely." In its discussion of possible consequence of various scenarios occurring at some time in the future, the organization often categorizes as to the *likelihood* of an event's occurring. For example, in one IPCC report that includes projections of various environmental conditions over the next century, authors report that

It is *virtually certain* that there will be more frequent hot and fewer cold temperature extremes over most land areas on daily and seasonal timescales, as global mean surface temperature increases. It is *very likely* that heat waves will occur with a higher frequency and longer duration. Occasional cold winter extremes will continue to occur.

and

Warming is *likely* to exceed 2°C for RCP6.0 and RCP8.5 (*high confidence*), more likely than not to exceed 2°C for

RCP4.5 *(medium confidence)*, but *unlikely* to exceed 2°C for RCP2.6 *(medium confidence)*. ("Climate Change 2014. Synthesis Report. Summary for Policymakers," n.d., 10)

As shown in these statements, authors can also indicate how confident they are in their results, with phrases such as "high confidence" to "medium confidence" to "low confidence."

This system is extraordinarily important since it reflects and reinforces the fact that statements made about future climates always contains an element of uncertainty and cannot be taken as statements that are simply *true*. Those statements always range from very likely to very unlikely to be true.

It is within this system of "likeliness" that one can review many predictions about possible future consequences of climate change. The following sections deal with some, although certainly not all, of the areas within which predictions can be made.

Greenhouse Gas Emissions

Perhaps the most fundamental question one might ask about climate change in the future has to do with emission patterns. If carbon dioxide and other greenhouse gases (GHG) are responsible for increases in global temperature, and those increases are responsible for all other changes discussed here, then it is crucial to know what we can expect for GHG emissions in the future.

The problem here, of course, and as with all other predictions that follow here, is what kind of future will occur. Will humans realize the seriousness of the climate change threat and act quickly to reduce GHG emissions? Or will they decide the threat is overblown and that a "wait and see" attitude is the wisest path to follow? Or will they choose to adopt some type of "in-between" plan that includes elements of both of these positions?

The IPCC has developed a system for making predictions for GHG emissions that takes into account this dilemma. In

2000, the organization issued a special report on "Emissions Scenarios." They developed four possible "storylines" based on choices that the world might make over the next few decades. Each scenario was developed with consideration for a complex host of factors, such as rate of economic growth, world population characteristics, and the availability of new and more efficient technologies. The four scenarios are broadly defined as follows:

A1: The world experiences rapid economic growth, with the development of many new and more efficient technological systems. World population reaches a peak in about 2050 and then begins to decline. The A1 category is further subdivided into three more specific groups:

A1FI describes a society that continues to rely on fossil fuels as its primary source of energy.

A1T refers to a world in which renewable energy technologies have a significantly large role in the world's energy equation.

A1B represents a world in which some balance between these two positions has been reached.

A2: Economic growth and technological development occur more slowly than in the A1 scenario. Also, significant differences in these two categories occur in various regions of the world. In addition, population growth is not under control and continues to increase over the next century or so.

B1: Population peak and decline follow the pattern for scenario A1. But economic growth increases rapidly, with emphasis on service and information elements. Greater emphasis is placed on global solutions, rather than dependent on national and/or regional solutions.

B2: Development tends to follow the patterns in A2, except that greater emphasis is placed on the solution of problems at a local, rather than global, level. Changes in economics and technology also tend to take place more slowly and are more moderate than in the A2 scenario. (For a detailed discussion of this approach to future scenarios, see Nakićenović et al. 2000.)

In its Fifth Report (2013–2014), IPCC adopted a new set of scenarios, called Representative Concentration Pathways (RCP). These scenarios follow the general principle set out in the organization's earlier "Emissions Scenarios" publication but differ in the ways in which categories are defined. We use the older system in this book, mainly because it is somewhat easier to understand (to learn more about the RCP system, see Wayne 2013; "Climate Change 2014. Synthesis Report. Summary for Policymakers," n.d.; for a comparison of the two systems, see the latter reference, 15, 21).

With this background, one can follow the predictions made for GHG emissions over the next century. The rate of emission for carbon dioxide at the beginning of IPCC assessments (1990) was 6.2 gigatons of carbon equivalent per year (GtC). That would, of course, be the same for all scenarios. The predictions for two intermediary future years (2020 and 2060) and the final year of prediction (2100) are shown in table 2.2. Notice that GHG emissions increase for almost all gases under all four scenarios, although much more rapidly for some gases under some scenarios. These data are more readily understood by looking at graphs for the events (see, e.g., "Predictions" 2006; Houghton et al. 2001). These graphs also include predictions for A1B, A1Fl, and A1T scenarios, which take into consideration the possibility of dramatic cutbacks in anthropogenic release of GHGs.

An additional piece of data of some interest is the projected total accumulation of GHGs by the year 2100. Although the *annual* emission of carbon dioxide and other gases may have begun to diminish at that point, the *total amount* of a gas may continue to increase. The reason for this phenomenon, of course, is that carbon dioxide can have a large residence time in

Table 2.2   Predicted GHG Emissions for Selected Gases, 1990–2100

Scenario A1

Gas	Unit*	1990	2020	2060	2100
CO_2 Fossil and Industry	GtC	6.2	12.2	15.9	13.2
CO_2 Deforestation	GtC	1.1	0.7	–0.3	–0.6
CH_4 (methane)	Mt	322.2	438.8	463.6	300.6
N_2O (nitrous oxide)	Mt	6.3	7.7	6.4	4.0

Scenario A2

Gas	Unit*	1990	2020	2060	2100
CO_2 Fossil and Industry	GtC	6.2	10.9	18.2	28.8
CO_2 Deforestation	GtC	1.1	1.7	0.7	0.2
CH_4 (methane)	Mt	322.2	448.3	674.0	912.7
N_2O (nitrous oxide)	Mt	6.3	9.4	12.7	16.4

Scenario B1

Gas	Unit*	1990	2020	2060	2100
CO_2 Fossil and Industry	GtC	6.2	8.3	10.4	6.5
CO_2 Deforestation	GtC	1.1	1.3	0.7	1.4
CH_4 (methane)	Mt	322.2	395.8	444.8	378.8
N_2O (nitrous oxide)	Mt	6.3	8.1	8.8	8.0

Scenario B2

Gas	Unit*	1990	2020	2060	2100
CO_2 Fossil and Industry	GtC	6.2	8.9	11.6	13.7
CO_2 Deforestation	GtC	1.1	1.6	–0.2	–0.2
CH_4 (methane)	Mt	322.2	447.5	544.4	508.4
N_2O (nitrous oxide)	Mt	6.3	7.1	5.8	5.1

* GtC = gigatons of carbon equivalent per year
Mt = megatons

Source: Nakicenovic, Nebojša, and Rob Swart, et al. 2000. "Emissions Scenarios." Intergovernmental Panel on Climate Change. New York: Cambridge University Press, Table 6.3, 356–357. https://www.ipcc.ch/site/assets/uploads/2018/03 /emissions_scenarios-1.pdf. Accessed on September 3, 2019.

the atmosphere, that is, a long period of time before it returns to Earth in some form or another. Depending on the form in which it occurs, the residence time for carbon dioxide can be anywhere from 5 to 200 years or more (Gruber and Steenken 2018). All that means is that some of the carbon dioxide emitted into the atmosphere in 2020 will remain there until 2030, 2050, and even beyond. The *total* present at any one time may, therefore, continue to grow. Graphs that illustrate this pattern can be found at Houghton et al. (2001, Figure 18, 65) or at "Projections for Carbon Dioxide" (2012).

Global Temperatures

The next, logical projection one might expect to find is global temperatures between the present day and some future point, such as the year 2100. And, as one might also expect, temperature trends tend to follow GHG emission trends. As table 2.3 indicates, all six of the most recent scenarios result in temperature increases in the global surface air temperature by 2100. The smallest increase, 1.98°C, occurs with the most optimistic of scenarios, B1, with dramatic changes in economics, population, and energy sources. The largest increase, 4.49°C, comes with the "business-as-usual" model in practice today. Under any scenario at all, then, we should expect an increase in the planet's global temperature by anywhere from 3.56°F to 8.08°F by the end of this century.

In some ways, that may seem like a small amount. The temperature can change by that much in a single hour on Earth today without any especially noticeable effect. But a comparison with the last global warming period in history, the Medieval Warm Period (MWP) of about 900 to 1300 CE is instructive. Scholars have long disagreed about relative temperatures during the MWP and our modern world. But the most recent data suggest that current warming trends are greater today than they were during the WMP. Furthermore, the temperature anomaly observed today appears to be at least twice that present during the WMP. And to the extent that the climate scenarios

Table 2.3   Model Average Surface Air Temperature Change, 1750–2100 (°C)

Year	A1B	A1T	A1FI	A2	B1	B2	IS92A
1750–1990	0.33	0.33	0.33	0.33	0.33	0.33	0.34
1990	0.00	0.00	0.00	0.00	0.00	0.00	0.00
2000	0.16	0.16	0.16	0.16	0.16	0.16	0.15
2010	0.30	0.40	0.32	0.35	0.34	0.39	0.27
2020	0.52	0.71	0.55	0.50	0.55	0.66	0.43
2030	0.85	1.03	0.85	0.73	0.77	0.93	0.61
2040	1.26	1.41	1.27	1.06	0.98	1.18	0.80
2050	1.59	1.75	1.86	1.42	1.21	1.44	1.00
2060	1.97	2.04	2.50	1.85	1.44	1.69	1.26
2070	2.30	2.25	3.10	2.33	1.63	1.94	1.52
2080	2.56	2.41	3.64	2.81	1.79	2.20	1.79
2090	2.77	2.49	4.09	3.29	1.91	2.44	2.08
2100	2.95	2.54	4.49	3.79	1.98	2.69	2.38

*A scenario developed by IPCC in a 1992 report: "IPCC IS92 Scenario. 2019." Data Distribution Centre. https://sedac.ciesin.columbia.edu/ddc/is92/index.html. Accessed on September 4, 2019.

Source: Houghton, J. T., et al., eds. 2001. "Climate Change 2001: The Scientific Basis." Intergovernmental Panel for Climate Change. New York: Cambridge University Press, 824. https://www.ipcc.ch/site/assets/uploads/2018/03/WGI_TAR _full_report.pdf. Accessed on September 3, 2019.

summarized in table 2.3 are accurate, those anomalies may be at least ten times as great in the near future ("How Does the Medieval Warm Period Compare to Current Global Temperatures" 2015, citing Moberg et al. 2005; Zhou et al. 2011). It would appear that current warming trends hold the possibility for creating significant environmental effects on the planet, as well as possible important changes in human civilization.

Glaciers and Ice Sheets

Possibly the most underappreciated feature of Earth's surface is its glaciers and ice sheets. Our planet is truly "the blue planet" or "the water planet." About 71 percent of its surface is covered with water. Of this amount, 96.5 percent occurs in the oceans, with another 0.9 percent in the form of other saline resources.

That leaves only about 2.5 percent of Earth's water in the form that humans most commonly use, fresh water. And of that fresh water, 68.7 percent is tied up in glaciers and ice sheets ("Where Is Earth's Water?" n.d.; an *ice sheet* is a dome-shaped mass of ice covering more than 50,000 square meters [12 million acres] in size; a *glacier* is a thick mass of ice that forms over many years as the result of continuing snow falls). Thus, nearly 30 times as much of Earth's fresh water occurs in the form of snow and ice, compared to reservoirs with which we are more familiar: groundwater, lakes, streams, and other forms of surface water. It's difficult to imagine what might happen if all or even part of that frozen water were to melt because of rising global temperatures. What are the chances that such a change might occur?

IPCC data on glaciers and ice sheets are generally unreliable prior to recent decades. During the period of the five reports, however, these data are abundant, with trends that are very clear. In AR5, for example, authors concluded that

> Most components of the cryosphere (glaciers, ice sheets, and floating ice shelves; sea, lake, and river ice; permafrost and snow) have undergone significant changes during recent decades (high confidence), related to climatic forcing (high confidence; WGI AR5 Chapter 4). (Field and Barros 2014, 987)

They also conclude that it is "*likely*" that these changes are the result of anthropogenic activities. (For a summary of observed changes in glaciers and ice sheets, see table 2.4.)

The two major results of glacial and ice sheet shrinkage are an increase in sea levels and an uplift of land on which the ice is located. In the former case, glacial and ice sheet melting is estimated to be responsible for about one-third of sea-level rise that has been documented to be a result of anthropogenic factors so far. Changes in the Antarctic ice sheet has also been documented, although there is a lower level of confidence that this change has an anthropogenic basis (Field and Barros 2014, 190).

Table 2.4 Observed Impacts of Climate Change Reported since AR4 on Mountains, Snow, and Ice, over the Past Several Decades, across Major World Regions, with Descriptors

Region	Snow and Ice	Confidence in Detection	Role of Climate	Climate Driver	Confidence in Attribution*
Africa	Retreat of tropical highland glaciers in East Africa	Very high	Major	Warming, drying	High
Europe	Retreat of Alpine, Scandinavian, and Icelandic glaciers	Very high	Major	Warming	High
Asia	Permafrost degradation in Siberia, Central Asia, and the Tibetan Plateau	High	Major	Warming	High
	Shrinking mountain glaciers across most of Asia	High	Major	Warming	Medium
Australasia	Substantial reduction in ice and glacier ice volume in New Zealand	High	Major	Warming	Medium
	Significant decline in late-season snow depth at three out of four alpine sites in Australia 1957–2002	High	Major	Warming	Medium
North America	Shrinkage of glaciers across western and northern North America	High	Major	Warming	High
	Decreasing amount of water in spring snowpack in western North America 1960–2002	High	Major	Warming	High

(continued)

Table 2.4   *(continued)*

Region	Snow and Ice	Confidence in Detection	Role of Climate	Climate Driver	Confidence in Attribution*
South and Central America	Shrinkage of Andean glacier	High	Major	Warming	High
Polar Regions	Decreasing Arctic sea ice cover in summer	Very high	Major	Air and ocean warming; change in ocean circulation	High
	Reduction in ice volume in Arctic glaciers	Very high	Major	Warming	High
	Decreasing snow cover across the Arctic	High	Major	Warming	Medium
	Widespread permafrost degradation, especially in the southern Arctic	High	Major	Warming	High
	Ice mass loss along coastal Antarctica	Medium	Major	Warming	Medium

*Attribution refers to possible links between human activity and observed natural events.
References for each event are omitted from this table.

Source: Field, Christopher B., and Vicente R. Barros, eds. 2014. "Climate Change 2014. Impacts, Adaptation, and Vulnerability. Part A: Global and Sectoral Aspects." Working Group II Contribution to the Fifth Assessment Report of the Intergovernmental Panel on Climate Change. New York: Cambridge University Press, Table 18.5, 1003. https://www.ipcc.ch/site/assets/uploads/2018/02/WGIIAR5-PartA_FINAL.pdf. Accessed on September 5, 2019.

All projections for the future of glaciers and ice sheets suggest that global warming will result in a continued loss in size. One basis for that projection is that all glaciers and ice sheets are too large for equilibrium with the present climate. That is, global temperatures have already increased to the point that the rate at which new snow and ice is added to a glacier or ice sheet during the winter is less than the rate at which ice and snow is lost during warmer seasons. Given those temperatures constraints, then, there appears to be no way of altering the reduction in size of glaciers and ice sheets in the foreseeable future (Field and Barros 2014, 243).

Loss of glacial ice and ice sheet over the past few decades has produced some of the most dramatic videos and images, as well as the most startling stories of changes resulting from these events during the period (see, e.g., Borunda 2019; Smith 2014). Possibly the most striking evidence of snow and ice loss has occurred in the small nation of Iceland. As its name suggests, snow and ice are an integral part of the country. Just over 10 percent of its land area is covered with glaciers. Visits to those glaciers by foreign tourists is a major source of income for Icelanders.

At least partly for this reason, the loss of glacial ice over the last few decades has been especially noticeable . . . and especially troubling. Those concerns reached a peak in August 2019, when one of the country's most notable glaciers, Okjökull (commonly known simply as "Ok") disappeared entirely. The event was memorialized at a well-attended ceremony on the site of the former glacier, along with the dedication of a plaque reminding future visitors of the circumstances that led to the glacier's disappearance (Magnason 2019).

One of the interesting phenomena associated with the loss of glaciers in Iceland is that land areas formerly covered by ice have begun to rise in elevation. The loss of massive amounts of ice have apparently caused the underlying ground to rebound by about a centimeter (half an inch) per year (Ástvaldsson 2019).

Oceans

In September 2019, the IPCC issued a special report on the effects of climate change on the oceans and the cryosphere (the regions of Earth covered by snow and ice). The report contained devastating news about these effects and dire warnings as to future changes to be expected without immediate action to deal with the problem. The report began by summarizing the role of the hydrosphere in Earth's environment and the lives of every human on the planet. It made clear that the changes observed in the oceans and cryosphere already recorded affected not a relatively small number of people living on or near the oceans. Instead, humans depend so heavily on the planet's water resources, that even communities located hundreds of miles from the oceans should expect measurable and damaging effects from climate change. In purely economic terms, IPCC predicted that declines in ocean health and services were projected to cost the global economy $428 billion per year by 2050, and $1.979 trillion per year by 2100 (Abram et al. 2019, 1–6).

One of the most obvious conclusions of the special report had to do with individuals living along the coasts of oceans. It noted that about 28 percent of the world's population could be considered to be living in "coastal areas," nearly half of whom lived on land less than 10 meters (30 feet) above sea level. These individuals and communities are at direct risk from the rise in sea level as the result of global warming. (All data in this section from Abram et al. 2019, 1–3 to 1–5. The Summary for Policymakers portion of this report contains more than 100 specific observations and impacts that should be of concern both to decision makers and to the general public. Those items clearly summarize the major findings of the report for the average reader.)

Numerous warming effects on the oceans and cryosphere have already been recorded, including ocean warming (virtually certain), increase in ocean acidity (high confidence), rise in sea level (high confidence), shrinking and loss of mass from glaciers and ice sheets (high confidence), loss of ice mass from the Arctic (very high confidence), and increased melting of permafrost (high confidence).

With regard to future overall prospects, the report noted that, among the various projections of future climate change, the current situation corresponds most closely with the most pessimistic of these scenarios. That is, researchers have outlined a variety of possible outcomes for future climate changes depending on the degree to which humans take action to reduce the release of greenhouse gases into the atmosphere. With aggressive actions, long-term results would probably be "not so bad," while little or no action would result in "serious consequences." Data contained in the current ocean and cryosphere report indicates that the world is on the latter of these two tracks.

Accumulating evidence also points to the dangers of reaching a *tipping point*, a point in history beyond which conditions continue to worsen with little hope that they can be reversed. The current report indicates that Earth has already reached, or is close to reaching, the tipping point for oceans and the cryosphere. That is, the changes that have already occurred in the environment—loss of glaciers and ice sheets, warming of the oceans, increased acidification, and rising sea levels, for example—are probably no longer reversible within the long stretch of human history. It may not be possible to reverse or even stop the further development of these changes no matter what actions humans may take.

The need for local and regional action to deal with ocean and cryosphere changes was an important theme of the report. It pointed out that, even if action on a global level does or does not occur, communities adjacent to the oceans can and should take actions to protect their own geographic, social, economic, and environmental concerns. Such actions, the report said, will be of value not only because they deal with specific local problems caused by climate change but also because they are relatively inexpensive, with few risks to the community and possible benefits to the preservation of local biodiversity.

Current findings about ocean and cryosphere changes and possible directions also emphasizes the importance of aggressive action on management and governance of possible responses to the problem. When talking about a global problem, such as the oceans, the issues and consequences involved for different communities, occupations, and regions are so great that more effective international and regional systems are needed to deal with such problems. Individuals and organizations with philosophies, goals, and approaches that are otherwise drastically different from each other must find more effective means of meeting, discussing, and taking action on this global problem.

Extreme Weather

Predicting future weather patterns, let alone future climate patterns, is a complex problem. So many factors are involved in the evolution of a storm that it is usually difficult to determine how much each of many factors may contribute to the development and intensity of a storm. A large amount of research has been conducted on the potential effects of climate change on hurricanes, tornadoes, and other types of severe storms, such as winter snow storms. The evidence adduced from that research is somewhat contradictory, but in general, it appears that the number and intensity of winter snow storms has been increasing in most parts of the United States since at least 1980 and is likely to increase even more in coming decades ("Climate Change Impacts in the United States" 2014). The reason for these changes appears to be a greater availability of the energy needed to produce such storms, primarily increases in atmospheric temperatures and wind shear factors (Trapp et al. 2007; also see Seeley and Romps 2015). Some research has begun to show an interesting side-effect to this future. It appears that the seemingly "minor" difference between a temperature increase of 1.5°C and 2.0°C can have major effects on the number and intensity of severe storms. This line of research suggests that the target for climate control suggested by many observers (2.0°C) may actually be too high (Barcikowska 2018).

The effect of climate change on hurricanes has also been studied in some detail. Given some inconsistencies in research results, there seems to be common agreement that there may be fewer hurricanes in the future, but those that do develop are likely to travel more slowly and be more severe than those in the past (Kossin 2018). The simplistic reason for this trend is that ocean water is the source of energy needed in the development of a hurricane. The warmer the seawater, the greater the amount of water evaporation, the more heat energy there is released from the evaporation, and the more intense the hurricane that develops. Abundant evidence dating back to the 1850s indicates that seawater has been warming at an accelerating pace from that period to the present day. The decadal temperature decreased from about 0.2°C in 1850–1860 to its lowest point in the period, –0.6°C in 1910–1920, before increasing to +0.4°C in 2018 ("Extended Reconstructed Sea Surface Temperature (ERSST) v4" 2018). According to one expert in the field, the magnitude of future hurricanes is likely to reflect the general rule that each degree (Fahrenheit) in ocean warming produces a 10 mile per hour increase in wind speeds in a hurricane. Since the oceans have already warmed (in 2019) by nearly 1.6°F over the 20th-century average, this new information may be of particular importance in predicting the characteristics of future hurricanes ("Global Climate Report—May 2018" 2018).

One of the interesting consequences of future hurricane behavior may be the need for a new intensity category. Currently, hurricanes are rated on a 1 to 5 basis on the Saffir-Simpson scale, with the upper limit on that scale (Category 5) being "157 mph or more." But the recorded speed for two 2017 hurricanes, Irma and Maria, already reached the upper level of the Category, about 180 mph. This trend suggests to at least a few observers that a new Category 6, or even Category 7, might need to be added to the scale to describe the worst future hurricanes (Fleshler 2018).

Based on the results of hurricane research, one might expect climatologists to have some ideas about the possible effects of climate change on the formation and intensity of tornadoes. In fact, such is not the case. Given that our understanding of tornadoes is already quite limited, it perhaps should not be surprising that very little is known about the way global warming might affect the genesis of tornadoes. In commenting on his recent study of tornado patterns over the past half century, Columbia University's Michael Tippett said that "[t]he fact that we don't see the presently understood meteorological signature of global warming in changing outbreak statistics [about tornadoes] leaves two possibilities: Either the recent increases are not due to a warming climate, or a warming climate has implications for tornado activity that we don't understand. This is an unexpected finding" (Evarts 2016; for the study cited, see Tippett, Lepore, and Cohen 2016). (Interestingly enough, the only tornado feature that so far has been associated with climate change is new pathways across the United States for such events, see Agee et al. 2016).

Some of the most interesting findings of the effect of climate change on extreme weather patterns comes from the emerging field of *attribution science*, also called *probabilistic event attribution*. This field of study attempts to estimate the extent to which anthropogenic climate change contributes to the severity of events such as extreme weather and forest fires. Attribution research usually depends on one of two approaches. One can actually study individual events and attempt to determine the change in the probability of that event's occurring or computer models can be used to calculate the difference between some natural events and that event as modified by human activities affects its probability.

The influence of human activities on the severity of certain natural events can no longer be denied. Each year, additional studies are reported demonstrating that the character of extreme storms, severe forest fires, or other disastrous events can be related directly to human activities. A good summary of this research can be found in annual reports published since 2011 by the American Meteorological Society ("Explaining Extreme Events" [annual report]; for full discussions of the science of attribution, see Burger, Horton, and Wentz 2020; National Academies of Sciences, Engineering, and Medicine 2016).

Land Use

Climate change affects land use by humans on Earth in a variety of ways. It may lead to a reduction in the amount of water available in a region, leading to desertification. It can also lead to degradation of land through erosion, loss of vegetation, damage by wildfires, and loss of or damage to permafrost. Climate change can also have significant effects on food security, that is, the amount, quality, and availability of foods needed to sustain human life. Threats to food resources come not only from a decline in the availability of land for agriculture but also from a reduction in the quantity and quality of food products grown on available lands.

In August 2019, IPCC published a summary report describing the land resources currently available on Earth, potential threats to those resources, the effects of those threats to human life, and the magnitude of each factor now and in the foreseeable future. The IPCC report summarized the risks to land use across a range of global mean surface temperatures (GMST), from 1°C to 3°C. It categorized the risks as ranging from "Very high probability of severe impacts/risks transition and the presence of significant irreversibility or the persistence of climate-related hazards, combined with limited ability to adapt due to the nature of the hazard or impacts/risks" to "Impacts/risks are undetectable" (Arneth et al. 2019, 13).

The report concludes that the risk posed to almost all resources with a less than 1°C rise in GMST would be essentially undetectable and of no consequence to the environment or human civilization. The report assigns a "high" level of confidence in this finding. The only exceptions to that finding involve permafrost resources and tropical crop yields, where consequential damage begins to occur at about 0.5°C increase in GMST. This situation changes with increases of greater than 1°C where "moderate" levels of harm are expected to occur for all listed land resources. ("Moderate" confidence for all data except for permafrost, which remains at "high" confidence.) Risk factors begin to climb to a level of "moderate" for nearly all resources as the increase in GMST exceeds 2°C. At this point, some resources, such as wildfire damage and food supplies, are at greater risk with global warming than are others. At temperature increases of greater than 3.5°C, all resources with the exception of soil erosion are classified as being at "very high" risk for serious damage (with a "moderate" level of confidence for all).

The conclusion that one can draw from these data is that the planet's land resources can withstand global temperature increases of less than 1°C with no noticeable effects. Beyond that point, however, each additional degree Celsius of increase brings with it a rapidly increasing level of concern. Beyond temperature increases of more than 3°C more than half of all land resources would be classified as being at "very high" risk for serious damage (Arneth et al. 2019, 13). These data are region-specific and represent global averages. Thus, loss of food supplies (or some other resource) might be more severe in Africa and less severe in North America. For reference purposes, the current global temperature anomaly is 0.8°C, up from 0.53°C in 2001 ("Global Temperature" 2019).

Wildfires

The first two decades of the 21st century have been marked by a seemingly unusually large number of forest fires of uncharacteristically severe intensity ("Facts and Statistics: Wildfires" n.d.). Experts have taken a variety of views as to why this change has occurred. On the one hand, there continues to be a historic debate about the role of fire management in preventing forest fires. Should forestry and fire entities use techniques such as clear cutting, brush removal, thinning, controlled burns, waste removal, and other methods for trying to prevent forest fires in the first place? Or are there natural forces that make fire *control* the more logical procedure than fire *prevention*? This debate has grown in intensity in the last few decades with the prediction and partial confirmation that climate changes have measurable effects on the number and intensity of forest fires. One particularly memorable expression of this debate occurred in November 2018 when a wildfire essentially destroyed the town of Paradise, California. In responding to this event, President Donald Trump said that, while climate change might have factored into that event "a little bit," the primary cause for the disaster was poor forest management. He pointed to the nation of Finland which, although heavily forested, experienced relatively few forest fires. He recommended that the United States follow the Finnish example and do more "raking and cleaning and doing things" (Kingsley 2018). At one point, Trump even considered ordering the Federal Emergency Management Agency (FEMA) to withhold future payments to California for such wildfires (Moon 2019).

So, what is the role of climate change, if any, in the increase in wildfire occurrence in the United States and other parts of the world? Attribution scientists have been especially interested in discovering what quantity and aspects of wildfires can be attributed to climate change, rather than to natural phenomena.

One of the most frequently cited studies on this topic was carried out by University of Idaho researcher John T. Abatzoglou and A. Park Williams of the Columbia University Lamont-Doherty Earth Observatory in 2016. They developed models for the contribution of anthropogenic activities to forest fires in the western United States from 2000 to 2015 and from 1979 to 2015. The found that human activities were responsible for an increase in temperature and a decrease in water vapor pressure resulting in more than 75 percent of the region to be at risk for wildfires during the period from 2000 to 2015. The increase in aridity from 1979 to 2015 resulting from human activities was about 55 percent. The sum total of these effects, they said, was an additional 4.2 million hectares (10.4 million acres) of forest lost to wildfires. Abatzoglou and Williams conclude that "[t]he growing ACC [anthropogenic climate change] influence on fuel aridity is projected to increasingly promote wildfire potential across western US forests in the coming decades and pose threats to ecosystems, the carbon budget, human health, and fire suppression budgets" (2016, 11774).

Other studies have produced similar, and sometimes even more extreme, results. For example, climate change researchers in British Columbia, Canada, modeled wildfires in the province during the 2017 fire season. A record 1.2 million hectares (3.0 million acres) burned during that season. How much of that loss, the researchers asked, was the result of human activity rather than natural causes? They found that there is a 95 percent probability that human activities increased the burned area by a factor of somewhere between 7 and 10 times. The researchers ended their report with the prediction that "[a]s the climate continues to warm, we can expect that extreme wildfire seasons like 2017 in BC will become more likely in the future" and that "other regions of the globe [will] experience strong relationships between climate and wildfires and with strong warming trends will likely also see increases in the likelihood of extreme wildfire seasons in the future" (Kirchmeier-Young et al. 2019, 8; for additional examples of this line of research, see Partain et al. 2016; Tett et al. 2018; Westerling et al. 2006; Williams et al. 2019; Yoon et al. 2015; for a good general summary of this topic, see "What Do We Know about Wildfire Attribution and Climate Change?" 2018)

Human Health

Climate change is expected to affect human health in a variety of ways. The details about these impacts are not yet entirely clear. For one thing, they are likely to differ substantially based on the region about which one is concerned. For example, developing nations are likely to experience greater effects in a number of fields at least partly because they begin with lower standards of living, lack of adequate public health facilities, shortage of funds for dealing with disease, and similar factors.

Thus, increased threat from a tropical disease as a result of climate change is likely to have a much more serious effect on residents of southern Africa than those of North America.

Second, relatively few studies have been conducted on the health effects of climate change compared to other effects, at least in part because of the time scale involved. One can measure the number of individuals killed in a severe storm that may be worsened by climate factors, but tracking an increase in illnesses and deaths because of communicable diseases, which often develop over weeks or months, many be more difficult to track. Nonetheless, some studies of this type have been completed, and others are currently being conducted. Experts now believe that climate change can affect human health in at least five major ways: temperature, extreme events, vector-borne diseases, food- and water-borne diseases, and food security and nutrition. The first two of these categories are sometimes classified as *direct* effects of climate change. For example, individuals who are killed in a hurricane die identifiably and immediately as a result of that event. Other categories are considered to be *indirect* effects because some intermediary steps may occur between a climate event and a health condition. For example, vector-borne diseases develop because of an increase in the number of insects or other disease agents in an area, an event that occurs because of changes in environmental conditions in that area.

Temperature

One hundred eight individuals died from heat in the United States in 2018. That number was just greater than the 10-year average for heat-related deaths (104) and less than the 30-year average (136). By contrast, the number of people who died because of extreme cold was 36, 32, and 30 for the three time categories (Benscoter 2019). The very numbers themselves point to one of the problems in predicting the effects of climate change. An increase or decrease of two or three might fall within the range of error for these measurements, so it would be difficult to determine the effects of climate change. One study has attempted to determine this effect using computer models of climate change and mortality rates. It found that increases in death rates would vary dramatically for various parts of the world. In some countries (mostly tropical nations), the death rate might increase by as much as 2,000 percent (Colombia). Rates greater than 775 percent would not be unusual. In other countries, death rates might increase by much less, with European nations among the most prominent of this group (Guo et al. 2018).

The effects of climate change on deaths due to extreme cold are expected to be just the opposite as those for extreme heat. After all, as the climate warms, the frequency of extreme cold days is like to decrease, resulting in fewer deaths from extreme cold. One summary on this issue has reported that "the reduction in premature deaths from cold are expected to be smaller than the increase in deaths from heat in the United States." In any case, the increase in heat-related deaths is expected to more than compensate for the decrease in cold-related deaths (Crimmins et al. 2016, section 2.7. U.S. data only; for worldwide estimates, see Gasparrini et al. 2017).

Extreme Events

Human health is affected both directly and indirectly by extreme weather events, such as tornadoes and hurricanes. In some cases, people are killed and severely injured from such events. For example, tornadoes cause an average of about 80 deaths and 1,500 injuries in the United States every year. Comparable numbers for hurricanes are 17 deaths and 60 injuries. Thunderstorms cause deaths and injuries in a variety of ways, such as by lightning and flooding. Each year, an average of about 55 people are killed by lightning and 70 by flooding (McNeill, n.d.). As the severity of severe storms increases, the number of individuals killed or injured is expected to increase.

People are also killed and injured indirectly as the result of severe storms. Some examples include the loss of access to food and water, which may cause nutritional or disease-related problems; damage to infrastructure, which may make it more difficult to transport the sick and injured to hospitals and other medical facilities; the loss of communication, making it difficult to send and receive information about the location and condition of injured individuals; and the creation or worsening of mental health problems, such as depression and post-traumatic stress disorder ("Climate Impacts on Human Health" 2017). Overall, the IPCC has predicted that an increase in global warming is *very likely* to result in a variety of harms to human health, although the specific magnitude of those numbers is not yet known (Smith et al. 2014, 722).

Vector-Borne Diseases

Vector-borne diseases (VBDs) are human illnesses caused by parasites, viruses, and bacteria that are transmitted by a variety of carrier organisms, such as mosquitoes, sandflies, triatomine bugs, blackflies, ticks, tsetse flies, mites, snails, and lice. Some of the best-known VBDs are dengue fever, hemorrhagic fever, Lyme disease, malaria, plague, and tick-borne encephalitis. These diseases may be transmitted by human-to-human contact, by way of one of these vectors, or by animal-to-human transmission. For example, plague is transmitted from rats to humans by means of fleas that have become infected by feeding on their host rats.

Research on the effects of climate change on VBDs is complicated for a number of reasons, one of the most important of which is geography. Most organisms involved in the transmission of VBDs live most successfully in warm humid environments, such as those found in southern Africa or the Indian subcontinent. As global temperatures rise, those conditions are likely to become more common in temperature and even arctic zones, thus increasing the likelihood of VBDs in areas where they are currently less likely to be found. On the other hand, modern medicine now has a variety of relatively effective drug treatments and other preventative measures for reducing the incidence of VBDs everywhere in the world.

A very large amount of research has now been done on the possible effects of a changing climate on VBDs. At least some of this research has been based on efforts to find attribution effects that will indicate the effects of projected climate changes on the spread and increase of VBDs in various parts of the world. Although some differences of opinion remain, the general consensus appears to be that a warmer climate will increase the occurrence of VBDs not only in tropical regions, where they tend to be most common, but also in temperate and eventually arctic regions. In the United States, the VBDs currently of greatest concern are chikungunya virus, dengue virus, Lyme disease, and West Nile virus. Existing research indicates that all four of these diseases have become more widespread in the United States over the last decade, although the role of climate change in that increase has not yet been well clarified (Caminade, McIntyre, and Jones 2018; Crimmins et al. 2016, ch. 5).

Food- and Water-Borne Diseases

Food- and water-borne diseases are illnesses caused by the transmission of bacteria, viruses, and parasites by way of contaminated food products (*food poisoning*) or contaminated water. Among the most common organisms involved this mechanism are bacteria, such as *Salmonella, Campylobacter, Clostridium, Listeria, Vibrio,* and *Yersinia*; viruses, such as norovirus, rotavirus, and hepatitis A; and parasites, such as *Toxoplasma gondii, Giardia duodenalis, Entamoeba histolytica,* and *Cryptosporidium parvum.*

As with VBDs, the possible effects of climate change on food- and water-borne diseases are difficult to predict because so many other factors are involved. Some preliminary research suggests that such diseases are likely to spread geographically as the global temperature warms. As an example, one study found that a warmer warm season in Africa may extend the period during which food- and water-borne diseases are most common, and hence increase the number of cases that might be expected. Another study found that warmer weather in the region surrounding Chicago might increase the overflow of sewage, and hence the spread of food- and water-borne pathogens, by anywhere from 50 to 120 percent by 2100 (Smith et al. 2014, 727). In the United States, current projections call for

> Increases in water temperatures associated with climate change [that] will alter the seasonal windows of growth and the geographic range of suitable habitat for freshwater toxin-producing harmful algae [*Very Likely, High Confidence*], certain naturally occurring Vibrio bacteria [*Very Likely, Medium Confidence*], and marine toxin-producing harmful algae [*Likely, Medium Confidence*]. These changes will increase the risk of exposure to waterborne pathogens and algal toxins that can cause a variety of illnesses [*Medium Confidence*]. (Trtanj et al. 2016, 157)

Food Security and Nutrition

The term *food security* refers to the ability of individuals to obtain and make use of sufficient amounts of nutritious food over an extended period of time. Researchers and climate change groups have devoted a substantial amount of time and energy attempting to determine possible effects of a global warming on food security for individuals in all parts of the world. Some projections might, on the face of it, seem likely. For example, in circumstances in which temperatures increase substantially or severe weather becomes more likely in agricultural areas, food production might be expected to decrease. As usual, projections for food security depend significantly on the computer model on which they are based which, in turn, is based on the scenario a researcher chooses for future population, technological, political, social, and other conditions. Some of the specific findings of research thus far are as follows:

- Climate change effects have already been detected on both land-based (*high confidence*) and aquatic (*medium confidence*) food products. Both positive and negative effects have been observed, but the latter are more common than the former.

- Crop yields tend to be very sensitive to changes in temperature (*high confidence*).

- Global temperatures approaching 4°C pose large risks to food security both globally and regionally (*high confidence*).

- Increases in global food prices by 2050 could range anywhere from 3 to 84 percent (*medium confidence*).

- Agronomic efforts to deal with these climate change effects may range from quite successful (with an increase of 15 to 18 percent in yields) to no effects to actual harm to crop production (*medium confidence*).

- Climate change is likely to have deleterious effects of aspects of food security extending beyond production, extending to and including storage, transportation, and consumption.

- Estimating the climatic effects of food security involves a complex mix of factors, including erosion, changes in weather patterns, changes in ecosystems, population growth, technological changes, and economic conditions. (Two excellent resources for further study of this problem are Brown et al. 2015; Porter et al. 2014.)

Ecosystems

Changes in the physical environment, such as alterations in temperature and rainfall, will inevitably produce changes in ecosystems. The term *ecosystem* refers to all the organisms present in a particular geographic region and the interactions that occur among those organisms. All of the organisms within any specific ecosystem have become adapted to the physical features of that ecosystem. For example, most fish species have evolved to live within a relatively narrow temperature range. In one study of North Atlantic marine species, for example, the tolerable range of temperatures for various species was about 25°C (45°F). A significant increase or decrease in ocean temperature for many of these species, then, would result in a die-off of populations or a decrease in rate of reproduction (Freitas et al. 2010, 3557).

Virtually all studies of climate change effects on ecosystems have produced similar results. In the first place, both land and marine species will tend to experience an upward drift, both in terms of geography and topography to cooler locations as their native habitats become warmer. This migration will be more likely for animals, which are able to move from one region to another, than for plants, which are anchored in a specific location. Thus, many animal species will be able to survive, albeit in a new location, while plant species are more likely to die off and become extinct.

For migratory species, it may be necessary to "reset the clock" for crucial life events, such as nesting and birthing of offspring. In one study, 28 migratory bird species have already altered their travel and nesting patterns to match warmer temperatures in the regions where they live ("Climate Impacts on Ecosystems" 2016). Studies of marine species have already found the type of changes described earlier, with those capable of doing so having already moved to cooler waters more similar to their original environments (Field and Barros 2014, 294–299).

Projections for the United States are at times startling. In one of the most comprehensive studies of climate change effects on ecosystems, the authors pointed out that the climate change effects on plants and animals of increasingly severe storms, flooding, and wildfires may be so severe that, as they predict, they may "cause many iconic species [to] disappear from regions where they have been prevalent or become extinct, altering some regions so much that their mix of plant and animal life will become almost unrecognizable" (Groffman et al. 2014, 196; resources of special importance include Field and Barros 2014; Grimm et al. 2013; Groffman et al. 2014; Malcolm and Pitelka 2000; McCarthy et al. 2001).

World Economics

One of the most common arguments used by climate change skeptics or deniers is that taking the actions needed to halt and/or reduce the damages caused by climate change would be too expensive. They would cause job loss and other economic damages that most countries could not afford. Given this argument, it is appropriate to ask what is actually known about the economic consequences to the United States and other nations if climate change solutions, such as carbon capture and carbon taxes, were implemented.

This topic has been the subject of several expansive and intensive surveys by economists. One of the most comprehensive of these studies examined the effect of two climate change scenarios on the value and costs of 22 sectors. The two scenarios were based on global temperature increases of 2.8°C and 4.5°C. The 22 sectors included categories such as air quality, labor, roads and bridges, coastal properties, urban drainage, electricity demand and supply, agriculture, coral reefs, and wildfire. The researchers' approach was to estimate the cost incurred within each sector, for each change scenario, that could be attributed to climate change. For example, in the area of coastal properties, they calculate the cost of abandonment of property, seawall construction, property elevation, and other actions taken to ameliorate the effects of climate change (Martinich and Crimmins 2019).

The primary conclusion of this study was that the overall cost of dealing with the effects of climate change under the more extreme scenario (4.5°C) would be $520 billion per year. Under the less severe scenario (2.8°C), the cost would be about almost half that amount, $300 billion per year. In either case, the economic costs of dealing with climate change would be substantial.

The sector most heavily affected by climate change under this model would be agriculture. Although agriculture is a relatively small fraction of the U.S. economy overall, it is a critical element of about a dozen states, and there is strong evidence that the predicted economic effects of climate change on agriculture are already being seen. Extreme rainfalls in the United States have increased by about 37 percent since the 1950s, resulting in severe flooding in many of the most heavily agriculture-dependent states (for more details, see Beach et al. 2015).

The sector next most vulnerable to climate change effects is infrastructure. Researchers calculated the costs of proactive maintenance and repairs to systems such as roads, bridges, and rail systems. Studies suggest that infrastructure losses would be experienced in essentially every part of the country (for a pictorial representation of regional effects of climate change on each of the 22 sectors, see Martinich and Crimmins 2019, fig. 1). A particularly interesting example of anticipated results of climate change on infrastructure involves a study of this effect on Internet infrastructure. Researchers found that 4,067 miles and 1,101 nodes of fiber optic cable would be damaged or lost by rising sea levels resulting from climate change. Those cables were made to be water-resistant (resistant in some degree to penetration of water, but not entirely) but not waterproof (completely safe from infiltration by water) (Durairajan, Barford, and Barford 2018; for an excellent summary of the economic effects of climate change, also see Tol 2014).

Data for the economic costs of climate change are not homogeneous worldwide. Some countries are apparently destined to confront far more serious problems than are others. One study on this issue attempted to place a specific dollar amount in economic costs for each ton of carbon dioxide (tCO_2) released. The leading country in this analysis is India, estimated to experience a cost of about US$90 per tCO_2, followed by the United States ($48/$tCO_2$), Saudi Arabia ($47/tCO_2), Brazil ($24/$tCO_2$), China ($24/tCO_2), and the United Arab Emirates ($24/$tCO_2$). Because these data reflect several factors, including geographical location and current climate change policies, some nations actually have negative $/$tCO_2$ values. They include Canada, Russia, and most northern European nations (Ricke et al. 2018; for better image view, see Nuccitelli 2018).

Climate Skeptics and Deniers

There has never been a time in history when climate change was *not* the subject of controversy. Until quite recently, those disputes took place between researchers in the field who argued about methods used in research, data collected, and the meaning of those data. Since about the 1960s, that situation has changed. As noted, for all practical purposes, there is unanimous agreement among scholars in the field about basic climate change facts: global temperatures are increasing; those temperature changes are not entirely the result of natural phenomena; the accumulation of carbon dioxide and other greenhouse gases has become, to a large extent, the result of human activities; continued accumulation of these gases and consequent temperature changes pose a serious threat to human civilization and the natural environment (https://iopscience.iop.org/article/10.1088/1748-9326/11/4/048002).

Although controversy over climate change continues in the early 2020s, the players in that debate come from different fields. On the one hand are experts in the field. On the other hand are industrialists, politicians, economists, and members of other professions and occupations, as well as some parts of the general public. This section will review some of the arguments posed by so-called *climate skeptics* and *climate deniers*, and responses provided by climatologists to these views.

To begin, here are some of the views expressed by climate skeptics and deniers:

"We are finding that the climate is not very sensitive to CO_2 and those kind of gases."
"The truth is, our climate system is so complex that we cannot predict its state even into next month."
John Christy, Professor of Atmospheric Science and
Director of the Earth System Science Center
at the University of Alabama Huntsville

"The supposed explanation that global warming is due to increasing atmospheric carbon dioxide from our burning of fossil fuels turns out to be based upon little more than circumstantial evidence."
Roy Spencer, principal research scientist for the
University of Alabama Huntsville;
Team Leader, Advanced Microwave
Scanning Radiometer, NASA's Aqua satellite

"In short, some parts of the IPCC process resembled a Soviet-style trial, in which the facts are predetermined, and ideological purity trumps technical and scientific rigor."
James Inhofe, U.S. Senator (R-OK); author of
The Greatest Hoax: How the Global Warming Conspiracy Threatens Your Future
(Washington, D.C.: WND Books 2012), 9

"The greatest hoax I think that has been around for many, many years if not hundreds of years has been this hoax on the environment and global warming."
Ron Paul, U.S. Representative (R/Lib-TX),
1976–1977, 1979–1985, 1997–2013

"There are a substantial number of scientists who have manipulated data so that they will have dollars rolling into their projects."
Rick Perry, Governor of Texas, 2000–2015,
U.S. Secretary of Energy, 2017–

"Sea level is rising at the unthreatening rate about a foot per century and decelerating."
Matt Ridley, 5th Viscount Ridley,
British journalist and businessman

(All quotes are taken from Cook 2019a. This source also offers responses to the objections raised here and by other skeptics and deniers. For another important source of quotes on climate change skepticism and denial, see "Global Warming Disinformation Database" 2019.)

The attitude of local council members, state representatives and senators, members of the U.S. Congress, and other decision-makers is often of special interest since these are the individuals who set policy and carry out actions dealing with climate change. A sample of denier quotes from members of the U.S. Congress follows.

Rep. Gary Palmer (R-AL): "The science shows that we haven't had a temperature increase in 17 or 18 years."

Rep. Tim Griffin (R-AR): "I am not convinced that the problem of global warming is what the scientists say it is."

Rep. Doug LaMalfa (R-CA): "The climate of the globe has been fluctuating since God created it."

Sen. Marco Rubio (R-FL): "I do not believe that human activity is causing these dramatic changes to our climate the way these scientists are portraying."

Rep. Rodney Davis (R-IL): ". . . global warming has stopped 16 years ago."

Rep. Steve King (R-IA): ". . . climate change 'is not proven, it's not science. It's more of a religion than a science.'"

Sen. Roy Blunt (R-MO): "There isn't any real science to say we are altering the climate path of the earth."

Rep. Steve Pearce (R-NM): "I googled this issue a couple of days ago, see that there are 31,000 scientists who say that human action is not causing the global warming at all."

Sen. Richard Burr (R-NC): "I have no clue [how much of climate change is attributable to human activity], and I don't think that science can prove it. . . . I certainly haven't seen anything that's conclusive, and anything that's claimed to be conclusive has proven to be somewhat sketchy."

Rep. Bob Gibbs (R-OH): "It is clear that science has not been able to document what is happening and if human activity is causing a problem or not."

Sen. James Lankford (R-OK): "This whole global warming myth will be exposed as what it really is—a way of control more than anything else."

Rep. John J. Duncan Jr (R-TN): ". . . global warming is "the greatest scam in history."

Rep. John Carter (R-TX): "Global warming is simply a chicken-little scheme to use mass media and government propaganda to convince the world that destruction of individual liberties and national sovereignty is necessary to save mankind, and that the unwashed masses would destroy themselves without the enlightened global dictatorship of these frauds."

Sen. Shelley Moore Capito (R-WV): "I don't necessarily think the climate's changing, no."

Sen. Ron Johnson (R-WI): "I absolutely do not believe in the science of man-caused climate change."

Rep. Mike Enzi (R-WY): "I barely made it back here because of a May snowstorm in Wyoming. They got 18 inches in Cheyenne. It's a little hard to convince Wyoming people there's global warming."

(All quotes taken from Germain et al. 2013. No members of the Democratic party are mentioned in this article. A similar article lists one Democratic officeholder as a climate denier. See Cranley 2019.)

Some analysts have attempted to develop systems for classifying the arguments put forward by climate skeptics and deniers. Those efforts have resulted in schemes that consist of anywhere from about 6 to nearly 20 "stages of denial," "myths," "misinformation," or other categories. Some examples of those categories are the following:

"Global warming is a natural phenomenon that has happened many times in the past."
    Global temperatures have changed over the millennia, but not as random, unexplainable events. Changes on the planet or the planet-sun system, such as its orientation in space or the occurrence of sunspots, have led to climate changes. But natural phenomena do not explain the pattern of climate change observed on Earth today.

"Carbon dioxide is only a trace gas and the amount added by human activities has no real effect on climate changes."

Research dating back more than a century confirms that carbon dioxide absorbs reflected heat from Earth's surface, thereby raising the average annual temperature of the planet and thereby affecting climate changes.

"Water is a far more abundant in Earth's atmosphere than is carbon dioxide, and climatologists routinely ignore its effect on climate change."

Water is not a forcing agent in climate change. It has a short residence time in the atmosphere, and any extra water vapor added to the atmosphere quickly returns to Earth's surface in the form of rain, snow, or some other type of precipitation.

"Some parts of the Earth are cooling, not warming. And unusually cold weather during winter is also occurring in some years."

These statements are "cherry-picking," selecting specific individual situations to prove a general point. In fact, all theories of climate change predict that some areas of the planet will be cooler than normal, a major reason that the term *global warming* has largely been replaced by the term *climate change*.

"Arctic sea levels are rising."

Very recent research confirms this observation. These data can be explained in a variety of ways, one of which is the increase in ocean water volume as glaciers in Antarctica and Greenland begin to melt. Local factors, such as changes in wind patterns, may also account for the observation.

"Global warming, to the extent that it exists, will be far more beneficial to humans and the natural environment than will current and previous cooler temperatures."

No one can say precisely what "more beneficial" means. If climate change were to occur over many centuries, that outcome is a possibility. But the rate at which the climate is changing currently provides living organisms, including humans, with relatively little time to adjust to new climatic conditions.

"Climatologists worldwide constitute a self-serving community that supports its members' views, often to their own benefit."

The well-known consensus among climatologists today about climate change reflects the abundance of research that confirms evidence such as that presented in IPCC reports, and not personal preferences or biases among researchers.

(These notes are based to a large extent on two extraordinary websites on climate skepticism and denial, Cook 2019b, and "How to Talk to a Climate Skeptic: Responses to the Most Common Skeptical Arguments on Global Warming" 2019.)

Industry Responds

News about the possibility of serious global warming and possible disastrous climate change was not well received by one very large segment of the world's economy: the fossil fuel industry. That fact should hardly be surprising. Scientists were pointing to carbon dioxide produced in the combustion of fossil fuels as a major, perhaps the major, cause of this new trend. Without providing a response to these claims, the fossil fuel industry potentially faced a destructive loss of income within a short period of time. They began a counteroffensive against the growing evidence of global warming, climate change, anthropogenic causes for such changes, and, in particular, on the combustion of fossil fuels as a major source of those changes.

One approach adopted by fossil fuel companies was to find or contact scientists who were willing to accept funding for research on climate change, research that might perhaps be expected to dispute the findings of most climatologists at the time. One classical example of this approach has been

Dr. Wei-Hock ("Willie") Soon. Soon holds a PhD in aerospace engineering from the University of South Carolina. Throughout his career, he has been affiliated with or received funding from several organizations associated with the fossil fuel industry, including ExxonMobil, the American Petroleum Institute, the Charles Koch Foundation, and the Atlanta-based electric utility, Southern Company. According to some reports, he has received more than $1 million from these organizations in support of his research ("Willie Soon" 2019). Among the many articles Soon wrote about climate change, the one for which he is perhaps best known is one that appeared in the journal *Climate Research* in 2003. In that article, Soon, along with co-author Sallie Baliunas, concluded that "[a]cross the world, many records reveal that the 20th century is probably not the warmest nor a uniquely extreme climatic period of the last millennium" (Soon and Baliunas 2003). Soon is by no stretch of the imagination the only researcher who has been accused of compromising his or her studies in an effort to misrepresent climatic patterns (see, e.g., "Climate Disinformation Database" 2019; "Holding Major Fossil Fuel Companies Accountable for Nearly 40 Years of Climate Deception and Harm," n.d.; Mulvey and Shulman 2015).

In their efforts to present an alternative view of climate change to the American public, fossil fuel companies sometimes developed joint programs of various kinds to achieve these goals. One of the best known of these plans was the so-called Global Climate Science Communications Plan. The plan was sponsored by the American Petroleum Institute and developed by representatives from several fossil fuel companies, including BP, ConocoPhillips, Chevron, ExxonMobil, and Shell. A report on the plan's development noted that "victory will be achieved when" a number of changes have occurred, including the acceptance by average citizens and the media of the "uncertainties" in climate science, the media provides more articles about viewpoints on climate science that challenge "the conventional wisdom," and those promoting the Kyoto Treaty "appear to be out of touch with reality" (Walker 1998).

A key element of the plan was to target children and young adults, as well as their teachers, on their understanding of climate change issues, especially from the standpoint of the fossil fuel industry. One way of approaching this effort was an attempt to partner with the National Science Teachers Association by providing curriculum materials and other information for use by teachers in their classrooms (Jervey 2015). A budget of about $2 million was allocated for operation of the program's first year, with continuing amounts anticipated for future years. Prospective sources of funding, according to the proposal, were the American Petroleum Institute, Business Round Table, Edison Electric Institute, Independent Petroleum Institute of America, and National Mining Association (Walker 1998).

Another approach adopted by fossil fuel companies for publicizing their own view of climate change was the creation of new organizations that, at least in some cases, were presented as "nonpartisan" or "independent" groups that conducted research on climate change and distributed the results of that research to the general public, politicians, educators, and other groups of individuals. One such organization was the Information Council on the Environment (ICE). ICE was founded in early 1991 by three coal and energy organizations, the National Coal Association, the Western Fuels Association, and Edison Electrical Institute. Its purpose was to develop an aggressive public relations campaign that would "[r]eposition global warming as a theory (not fact)" (Information Council for the Environment 1991, "Strategies"). The organization had a budget of about a half million dollars, to be spent on five newspaper ads, radio spots and radio commercials, a public relations tour, and a letter-writing campaign conducted among interested persons. The program failed to complete its proposed program and went out of business within a year because a copy of its internal documents was leaked to the general public and disavowed by at least one member of the founding committee (Information Council for the Environment 1991, n.d.).

A more successful group was the Global Climate Coalition (GCC). GCC was organized in 1989 in response to a perception among fossil fuel companies that they had been slow in responding to developments in climate science up to that time. Several companies expressed a *mea culpa* in allowing the discussion over climate change's being "taken over" by environmentalists at meetings such as the Montreal Protocol of 1987 and the first meeting of the Intergovernmental Panel on Climate Change in 1988. Although scientific advisors from a variety of fossil fuel companies were already acknowledging the reality of climate change and human involvement in that process, the industry decided that it needed to become more aggressive in presenting another side of the story (Revkin 2009).

GCC was originally established under the auspices of the National Association of Manufacturers and consisted of more than 40 corporations and trade associations, including Amoco, the American Forest and Paper Association, American Petroleum Institute, Chevron, Chrysler, Cyprus AMAX Minerals, Exxon, Ford, General Motors, Shell Oil, Texaco, and the U.S. Chamber of Commerce. At its peak, the organization is said to have been representing more than 230,000 discrete firms (Burton and Rampton 1997; Franz 1998, 13). It took a multi-pronged approach to doing battle with climate scientists. As with other skeptic and denier groups, it argued that findings implicating anthropogenic activities in current warming trends was overblown and even incorrect. In one of its backgrounder documents, it argued that "The GCC believes that the preponderance of the evidence indicates that most, if not all, of the observed warming is part of a natural warming trend which began approximately 400 years ago. If there is an anthropogenic component to this observed warming, the GCC believes that it must be very small and must be superimposed on a much larger natural warming trend" ("1996 Global Climate Coalition: An Overview and Attached Reports," n.d., 2).

One of the strategies developed by GCC was to have representatives of member organizations "infiltrate" the IPCC by offering to conduct research, offer papers, and serve in other positions through which they could influence IPCC statements and policy positions. Although not particularly successful in these efforts, they were apparently more so in dealing with the administration of President George W. Bush. When Bush decided to withdraw from negotiations over the Kyoto Treaty in 2001, a member of the State Department wrote GCC to confirm that the president's decision was "based on input from you" ("2001 State Department Briefing for Global Climate Coalition Meeting" n.d.).

As news of the modus operandi of climate denier groups such as ICE and GCC became better known, their effectiveness as spokespersons for the fossil fuel industry began to diminish. In 1996, British Petroleum withdrew from the group, followed by Royal Dutch Shell a year later, Ford Motor Company and DuPont in 1997, and Shell (U.S.) in 1998. By 2000, several other important members had also withdrawn from the group, including Daimler-Chrysler, General Motors, the Southern Company, and Texaco. Shortly after Bush's decision to leave the Kyoto Treaty in 2001, GCC itself discontinued its operations (Savage 2019; for a good overview of GCC's history and activities, see "Global Climate Coalition" n.d.; Franz 1998; "Global Climate Coalition: Climate Denial Legacy Follows Corporations" 2019).

Over the next two decades, more and more climate change deniers were changing their minds about their historic position on global warming. At least part of the reason for this change was the continued accumulation of evidence for climate change and its potential effects on the environment. As early as 2006, the U.S. oil firm of Exxon Mobil had begun to reverse its position on climate change. New CEO Rex Tillerson (later President Donald Trump's first Secretary of State) announced an several occasions that, as unpleasant as taxes may be, a tax on carbon was almost certainly needed to fight global warming. So-called *carbon taxes* had long been (and are still today in many instances) anathema to the fossil fuel industry (Polansky 2017; Schwartz 2016).

Exxon Mobil was only the first of many companies to change its mind about climate change and carbon taxes. One of the most stunning of those turn-arounds came in 2015 when six large European oil companies—BG Group, BP, Eni, Royal Dutch Shell, Statoil, and Total—wrote a letter to Christiana Figueres, Executive Secretary of the United Nations Framework Convention on Climate Change (UNFCCC), asking that action be taken to deal with climate change. Their major suggestion was the imposition of a carbon tax on the use of fossil fuels ("Paying for Carbon Letter" 2015).

Some critics have suggested that fossil fuel companies have been somewhat disingenuous in now supporting efforts to deal with climate change. They argue that very large corporations such as Exxon Mobil and Royal Dutch Shell must find ways of surviving in a world that has begun to reject fossil fuels as a major energy source. Recent reports have confirmed the suspicion that large fossil fuel companies tend to talk a supportive position for climate change reform but rarely act in ways dictated by such beliefs. In 2019, for example, the financial think tank Carbon Tracker found that large oil companies had invested at least $50 billion in new fossil fuel projects in the previous year. At the same time, all of the major companies had spent very little—in the case of ExxonMobile, less than one-fifth of 1 percent—on the development of renewable energy (Fletcher et al. 2018; Grant and Coffin 2019; Skuce 2012; Tabuchi 2019).

Interestingly enough, the "conversion" of fossil fuel companies has been mirrored to some extent by a corresponding change of heart by well-known and once-widely-quoted experts in the field and by many members of the general public (Calamia 2010; "Former Climate Change Deniers, What Changed Your Mind?" 2017).

Possible Solutions

Many individuals and organizations concerned about climate change have proposed a variety of methods for ameliorating or eventually ending the problem. These solutions run the gamut from personal and quite simple to national or international and sometimes quite complex.

What Can You Do?

Many of the suggestions made for dealing with climate change focus on an individual's lifestyle. In most cases, they rely on ways for reducing the use of fossil fuels for some activity or another. One of the best-known examples of such an instance occurred in September 2019, when Swedish teenager Greta Thunberg arrived in New York aboard a sailing yacht from Plymouth, England. Having been inspired to become a climate change activist in her high school science classes, Thunberg decided to avoid any fossil-fuel-burning mode of transportation, such as an airplane or cruise ship, to travel to the United States. She was simply putting her beliefs about climate change into her everyday life (Chappell and Chang 2019).

Thunberg's action illustrates just one way a person can reduce her or his carbon footprint. The term *carbon footprint* is often used to describe an individual or corporation's contribution to global warming and climate change. Some other suggestions that have been made include the following:

- Consume less and waste less. For decades in developed countries, individuals have been taught simple to "throw it away." This philosophy has led to the need (a profitable one for corporations) to continue making more and more new products. By making purchases with conservation (and, therefore, less use of fossil fuels) in mind, a person can significantly reduce her or his carbon footprint.

- Think of ways to use energy more efficiently. For example, buy energy-efficient light bulbs; wash clothes in cold or warm, rather than hot, water; unplug electrical devices when they are not in use; winterize your home to prevent heat loss during the coldest part of the year; and set your home thermostat a few degrees lower in the winter and higher in the summer.

- Choose efficient means of transportation. The automobile is probably the single most inefficient mode of transportation today. Wherever possible, use carpooling, public transportation, ride a bike or walk, or simply reduce nonessential travel, for example, by traveling shorter distances for vacations.

- Change your eating habits. One of the largest contributors to global warming today is the raising of animals for human foods. Recent data suggest that cows, sheep, and other ruminants produce about 160 million tons of greenhouse gases every year. By eating less meat, one important cause of global warming can be reduced by a significant amount.

- Wherever possible, replace fossil fuel systems in your home, such as a natural gas furnace, with alternative energy sources, such as solar devices or wind systems.

- Become politically active. Any number of organizations exist through which individuals can work to reduce climate change. Actions may include campaigning for politicians who share your view about global warming, participating in campaigns to make your community more climate change sensitive, and working in your own neighborhood or school to increase awareness of climate change issues (for more ideas, see Goodall 2010).

What Can Your Community Do?

The change that five people working together can bring about is much more than five times the amount they can produce working alone. That is, groups of people can often achieve goals that are far beyond those of single individuals. Here are some of the projects neighborhoods and communities can consider to reduce climate change.

- Develop areas and policies that encourage people to use more efficient means of transportation. This could include increasing the number of electrical-vehicle charging stations; improving the community's bus system; or developing roads, highways, and signaling systems that allow cars and trucks to move more efficiently through the community.

- Adopt building codes and take other actions to ensure that both public and private structures are as efficient as possible. Many older buildings were constructed without much thought to conserving heat, cooling, and electrical uses, but changes can be made to correct this problem.

- Find alternatives for parking lots, where people are encouraged to continue personal transportation while using valuable land.

- Promote tree-planting programs. Trees and other plant life are still one of the most essential components of a planetary system for removing carbon dioxide from the air safely and efficiently.

- Encourage local officials to develop a more aggressive climate change policy by modifying building codes, encouraging use of alternative energy, discouraging the use of private vehicles, improving education about global warming, and similar activities (also see Sisson, Barber, and Walker 2019).

What Can Your Country Do?

Some of the most contentious discussions about dealing with climate change have come not at the personal or local level, but at the national, regional, and international level. After all, climate change is an issue that affects every person, every community, and every nation in the world.

A U.S. city, county, or state that adopts a truly aggressive approach to reduce climate change, then, can have only limited, even if vital, effects worldwide. Dealing with *global* warming in a truly effective way, then, will probably occur only when most nations in the world can agree on the steps that they can take, and that they are willing to take, undertake. Over the dozens of climate change conferences, then, the debate has often been over the remedial steps that can be taken, and that all participants are willing to sign off on (for a list of such conferences, see "Events and Conferences" 2019).

The basic approach behind the most popular methods for dealing with climate change is called *carbon pricing*. That term refers to the practice of charging companies and other entities for the carbon (generally in the form of carbon dioxide) that they release to the atmosphere. It is based on the principle that an entity should be made financially responsible for the social, economic, and other costs of producing a material that causes damage to others. For example, companies that emit carbon dioxide to the atmosphere today pay nothing for the privilege of doing so. They get to "throw away" a waste product for which they have no use, at no cost to themselves. Going back as far as 1920, some economists have said that such a practice should not be allowed ("The Pigou Club" 2012). Entities that cause damage to the environment, human health, or other parts of human society should contribute financially to dealing with these problems. Carbon pricing is such an approach.

Carbon pricing can take a variety of forms. One such approach is *carbon trading* or *carbon emissions trading*. In this system, the amount of emissions produced by a company over some historical period is determined. A supervising agency, often a governmental agency, then issues a certain number of permits, or *allowances*, equal in value to those emissions. In the future, that company has one of three choices. It can continue to produce the same amount of emissions as it has in the past—that is, it simply retains the number of permits it was originally allotted. Or, it can decide to reduce its emission, thus having more permits than it now needs. That decision allows the company to sell its extra permits to some other company that wants to increase its level of emissions. And that is the original company's third option: it can decide that it wants to increase its level of emissions—but it can do so only by finding another company willing to sell some of its own permits.

By far the most popular form of carbon trading is a so-called *cap-and-trade* system. It is essentially identical to carbon trading except that it establishes at the outset a specific limit on the total amount of emissions allowed under the scheme. The total is equal to the sum of all emissions by all participants in the cap-and-trade system. That cap can be raised or lowered by the governing entity, but once decided upon, it sets the limits within which carbon trading can take place.

Cap-and-trade has many advantages as a method for reducing global warming and climate change. For one thing, the governing entity has the power to gradually reduce the cap over time, thus, for all practical purposes, reducing the amount of emissions released by companies to the atmosphere. The system also allows companies to "retire" permits if they choose to contribute to the control of climate change. That is, a company may own 100 permits, but use only 80 of those permits in its own activities. It can also decide not to sell the remaining 20 permits, thus reducing the amount of carbon emissions. Cap-and-trade also provides additional revenue for the governing entity. For example, suppose that a division of the U.S. government responsible for cap-and-trade conducts an auction to sell shares in the program. That auction can provide additional income to the governmental agency involved. Perhaps most attractive of all to many companies is that the cap-and-trade scheme is truly a market-based system. Except for the government's possible control of the cap itself, most of the activities that occur in the scheme involve bargaining among companies (Regoli 2019).

Cap-and-trade also has its disadvantages. Although it is essentially a market-based system, cap-and-trade ultimately depends on government-based caps, within which individual companies must operate. This provision may impose undue burdens on a company or cause it to reduce emissions even when that might be the best business practice. Some are concerned about a company's ability to cheat in the way it buys, sells, and uses the permits it owns. Also, on a global level, the task of simply setting up such a system to which all countries can agree is a daunting challenge (Regoli 2019).

Cap-and-trade systems already exist in many parts of the world and, in some cases, have been operating successfully for several years. Some examples include the Emission Trading Scheme in the European Union; the Regional Greenhouse Gas Initiative in ten Northeastern and Mid-Atlantic states (electrical generation only), the Western Climate Initiative in seven western states and four Canadian provinces, and the Midwestern Regional Greenhouse Gas Reduction Accord, consisting of six Midwestern states and the province of Manitoba; and similar systems in New Zealand, South Korea, Kazakhstan, and parts of China and Japan (for a complete list of countries and regions, see Ackva et al. 2018).

A second popular form of carbon pricing is the *carbon tax*. As the name suggests, a carbon tax is nothing more nor less than the fee paid by an entity for the carbon (again, usually carbon dioxide) that it releases to the atmosphere. The first country to impose a carbon tax was Finland, which adopted the policy in 1990. Since that time, nearly 30 other nations and other entities have adopted some form of carbon tax, including the countries of Australia, Chile, Costa Rica, Iceland, India, Ireland, Poland, Sweden, Switzerland, the United Kingdom, and Zimbabwe, as well as the provinces of Alberta, British Columbia, and Quebec in Canada, and the city of Boulder, Colorado. Several other nations and other governmental groups have been or are actively considering the use of carbon taxes to reduce carbon emissions ("Where Carbon Is Taxed" 2018).

The size of a carbon tax varies widely worldwide, with the highest tax being assessed in Sweden, US$131 per tCO_2e (tons of carbon-dioxide equivalent), followed by Switzerland ($86 per tCO_2e), Finland ($69/66 per tCO_2e, depending on source), and Norway ($52 per tCO_2e). Most countries charge less than $10 per tCO_2e, with the tax in Mexico and Poland set at less than $1 per tCO_2e ("Where Carbon Is Taxed" 2018; for an excellent overview of carbon taxes, see Ye 2013). Other forms of carbon pricing exist but tend to be somewhat less common (for more information on these systems, see World Bank, Ecofys, and Vivid Economics 2017).

Technological Solutions

The economic solutions to climate change reviewed here are unacceptable to many individuals, especially climate skeptics and deniers, because they impose a financial burden on businesses. They argue that the case for climate change is not strong enough for nations or other governmental units to impose heavy economic assessments on private companies. Instead, they may offer other types of solutions (if they suggest any at all) for dealing with climate change. Most of these suggestions involve technological approaches the offer a way to reduce the amount of carbon dioxide and other greenhouse gases in the atmosphere by physical or biological means. They are often classified as forms of *geoengineering*, large-scale interventions in Earth's natural systems to counteract climate change.

Geoengineering projects typically fall into one of two categories: solar radiation management (or solar radiation modification; SRM) or carbon-dioxide removal (CDR). A recent IPCC report lists four major types of SRM:

- stratospheric aerosol injection, in which particles are injected into the atmosphere that will reflect solar radiation, thus reducing the amount that reaches Earth's surface. The effect is similar to the one that occurs when a volcano erupts, ejecting reflective particles into the atmosphere;

- marine cloud brightening, which involves spraying salt crystals into marine clouds to make them more reflective;

- cirrus cloud thinning, which allows more long-wave radiation from Earth's surface to escape back into the atmosphere; and

- ground-based albedo modification, in which large portions of Earth's surface are brightened, such as changing the way farming is done, to reflect a large amount of heat from Earth's surface ("Global Warming of 1.5°C" 2016, Table 4.7, p. 348).

As the name suggests, CDR systems involve the capture of carbon dioxide from the atmosphere by some means and then storing it underground or in the oceans. These processes are often said to involve *negative emissions* because the carbon dioxide produced in a process is captured and stored underground or in the oceans. The most popular form of CDR now being considered is called bioenergy with carbon [dioxide] capture and storage (BECCS). In this system, carbon dioxide from the air is used for some natural process, such as growing plants. Those plants are then harvested and burned to produce energy. The carbon dioxide produced in this reaction is then captured and stored underground. The IPCC has acknowledged that this process might be a feasible way to remove some small fraction of atmospheric carbon dioxide in the future ("Global Warming of 1.5°C" 2016, 326). That conclusion has been disputed, however, by some experts in the field ("In-depth: Experts Assess the Feasibility of 'Negative Emissions'" 2016). The IPCC itself has also expressed some doubts as to the extent to which SRM and CDR technologies can make significant contributions to long-term efforts to control climate change ("Global Warming of 1.5°C" 2016, 316–317; for more detailed discussions of SRM and CDR, see National Research Council 2015a, 2015b; Santos 2019).

Conclusion

Since the late 20th century, several scientists have been warning the general public and politicians that humans have embarked on a "great experiment" of Earth's climate. Without having planned to conduct such an experiment, humans have actually been finding out what can happen to the climate by dumping ever-increasing amounts of carbon dioxide into the atmosphere every year. The problem is that that experiment has not been planned by anyone nor is there obvious way to bring it to an end.

We are, however, beginning to see some possible consequences of the experiment: Earth's temperature has begun to rise, storms are becoming more severe, glaciers and ice sheets are beginning to melt, and ecosystems are starting to change in significant ways. So, we can conclude that, yes, it seems quite likely that the climate *is* changing and human activities *are* responsible for these changes.

In spite of the growing evidence about climate change, however, differences of opinion continue to exist as to whether that change is real, whether humans are responsible in any ways, what can be done to reverse the process, whether it's worth the cost of moving on these options, and so on. So, in some ways, the results of this great experiment will not be known for decades, or even centuries. The troublesome fact about that scenario is that it is now too late, or approaching the "point of no return," to slow or stop climate change. Carbon dioxide is accumulating in the atmosphere, and it will be there for a long time to come. By the time humans begin to see the most serious consequences of climate change, it may well be too late to act to prevent those consequences.

So, in some ways, the fundamental climate change debate today is (1) do we spend the time and resources today to combat possible climate change and suffer possible economic and social costs to humans alive today or (2) do we focus on maintaining our prosperity and way of life today, taking the chance that purported serious effects of climate change are overblown and/or unrealistic. We probably can't have it both ways.

References

Abatzoglou, John T., and A. Park Williams. 2016. "Impact of Anthropogenic Climate Change on Wildfire Across Western US Forests." *Proceedings of the National Academy of Sciences of the United States of America* 113(42): 11770–11775. https://wildfiretoday.com/documents/Fires _Climate_Change.pdf. Accessed on September 8, 2019.

Abram, Nerilie, et al. 2019. "IPCC Special Report on
the Ocean and Cryosphere in a Changing Climate."
Intergovernmental Panel on Climate Change. https://
report.ipcc.ch/srocc/pdf/SROCC_FinalDraft_FullReport
.pdf. Accessed on September 25, 2019.

Ackva, Johannes, et al. 2018. "Emissions Trading Worldwide.
Executive Summary." International Carbon Action
Partnership. https://icapcarbonaction.com/en/?option=com
_attach&task=download&id=528. Accessed on September
16, 2019.

Agee, Ernest, et al. 2016. "Spatial Redistribution of U.S.
Tornado Activity between 1954 and 2013." *Journal of
Applied Meteorology and Climatology* 55(8): 1681–1697.
https://journals.ametsoc.org/doi/pdf/10.1175/JAMC-D-15
-0342.1. Accessed on September 6, 2019.

Arneth, Almut, et al. 2019. "IPCC Special Report on
Climate Change, Desertification, Land Degradation,
Sustainable Land Management, Food Security, and
Greenhouse Gas Fluxes in Terrestrial Ecosystems. Summary
for Policymakers." Intergovernmental Panel on Climate
Change. https://www.ipcc.ch/site/assets/uploads/2019/08
/Edited-SPM_Approved_Microsite_FINAL.pdf. Accessed
on September 6, 2019.

Ástvaldsson, Jóhann Páll. 2019. "Land Rising Due to Melting
Glaciers." Iceland Review. https://www.icelandreview.com
/news/land-rising-due-to-melting-glaciers/. Accessed on
September 5, 2019.

Barcikowska, Monika J., et al. 2018. "Euro-Atlantic Winter
Storminess and Precipitation Extremes under 1.5°C vs. 2°C
Warming Scenarios." *Earth Systems Dynamics* 9: 679–699.
doi:10.5194/esd-9-679-2018. https://www.earth-syst
-dynam.net/9/679/2018/esd-9-679-2018.pdf. Accessed on
September 6, 2019.

Beach, Robert H., et al. 2015. "Climate Change Impacts on
US Agriculture and Forestry: Benefits of Global Climate

Stabilization." *Environmental Research Letters* 10(9): 095004. doi:10.1088/1748-9326/10/9/095004. https://iopscience.iop.org/article/10.1088/1748-9326/10/9/095004/pdf. Accessed on September 15, 2019.

Benscoter, Jana. 2019. "Heat Stroke Tops List of Weather-Related Deaths." Penn Live. https://www.pennlive.com/news/2019/07/heat-stroke-tops-list-of-weather-related-deaths.html. Accessed on September 9, 2019.

Borunda, Alejandra. 2019. "Most Dire Projection of Sea-Level Rise Is a Little Less Likely, Reports Say." National Geographic. https://www.nationalgeographic.com/environment/2019/02/antarctic-greenland-ice-melt-less-bad/. Accessed on September 5, 2019.

Brown, Molly E., et al. 2015. "Climate Change, Global Food Security, and the U.S. Food System." U.S. Department of Agriculture. University Corporation for Atmospheric Research. National Center for Atmospheric Research. http://www.usda.gov/oce/climate_change/FoodSecurity2015Assessment/FullAssessment.pdf. Accessed on September 13, 2019.

Burger, Michael, Radley Horton, and Jessica Wentz. 2020. "The Law and Science of Climate Change Attribution." *Columbia Journal of Environmental Law*. In press. https://poseidon01.ssrn.com/delivery.php?ID=105078117027006081031085092095095064030064049034070049097092106112065110065095121095050000006042010051053025098026028072024093019080035020033099118106091066127028073056062046025025092090066123095096018000106125088075014010079118126088124003004105020092&EXT=pdf. Accessed on December 11, 2019.

Burton, Bob, and Sheldon Rampton. 1997. "Thinking Globally, Acting Vocally: The International Conspiracy to Overheat the Earth." *PR Watch* 4(4): 1–6. https://www.prwatch.org/files/pdfs/prwatch/prwv4n4.pdf. Accessed on September 1, 2019.

Calamia, Joseph. 2010. "Global Warming Dissenter Bjorn Lomborg (Sort of) Has a Change of Heart." http://blogs.discovermagazine.com/80beats/2010/08/31/global-warming-dissenter-bjorn-lomborg-sort-of-has-a-change-of-heart/#.XW1CqihKhEU. Accessed on September 2, 2019.

Caminade, Cyril, Marie McIntyre, and Anne E. Jones. 2018. "Impact of Recent and Future Climate Change on Vector Borne Diseases." *Annals of the New York Academy of Sciences* 1436(1): 157–173. doi:10.1111/nyas.13950. https://www.ncbi.nlm.nih.gov/pmc/articles/PMC6378404/. Accessed on September 13, 2019.

Chappell, Bill, and Alisa Chang. 2019. "Greta Thunberg to U.S.: 'You Have A Moral Responsibility' on Climate Change." NPR. https://www.npr.org/2019/09/13/760538254/greta-thunberg-to-u-s-you-have-a-moral-responsibility-on-climate-change. Accessed on September 16, 2019.

"Climate Change Impacts in the United States." 2014. U.S. Global Change Research Program. https://www.nrc.gov/docs/ML1412/ML14129A233.pdf. Accessed on September 6, 2019.

"Climate Change 2014. Synthesis Report. Summary for Policymakers." n.d. Intergovernmental Panel on Climate Change. https://www.ipcc.ch/site/assets/uploads/2018/02/AR5_SYR_FINAL_SPM.pdf. Accessed on September 4, 2019.

"Climate Disinformation Database." 2019. Desmog. https://www.desmogblog.com/global-warming-denier-database. Accessed on August 31, 2019.

"Climate Impacts on Ecosystems." 2016. Environmental Protection Agency. https://19january2017snapshot.epa.gov/climate-impacts/climate-impacts-ecosystems_.html. Accessed on September 14, 2019.

"Climate Impacts on Human Health." 2017. Environmental Protection Agency. https://19january2017snapshot.epa.gov /climate-impacts/climate-impacts-human-health_.html #ref1. Accessed on September 13, 2019.

Cook, John. 2019a. "Climate Misinformation by Source." https://skepticalscience.com/misinformers.php. Accessed on August 30, 2019.

Cook, John. 2019b. "Global Warming & Climate Change Myths." Skeptical Science. https://skepticalscience.com /argument.php. Accessed on August 30, 2019.

Cranley, Ellen. 2019. "These Are the 130 Current Members of Congress Who Have Doubted or Denied Climate Change." Business Insider. https://www.businessinsider .com/climate-change-and-republicans-congress-global -warming-2019-2. Accessed on August 31, 2019.

Crimmins, A. J., et al., eds. 2016. "The Impacts of Climate Change on Human Health in the United States: A Scientific Assessment." U.S. Global Change Research Program. doi:10.7930/J0R49NQX.

Durairajan, Ramakrishnan, Carol Barford, and Paul Barford. 2018. "Lights Out: Climate Change Risk to Internet Infrastructure." Conference Paper. doi:10.1145/3232755.3232775. http://pages.cs.wisc.edu /~pb/anrw18_final.pdf. Accessed on September 15, 2019.

"EPA Finalizes Affordable Clean Energy Rule, Ensuring Reliable, Diversified Energy Resources while Protecting our Environment." 2019. Environmental Protection Agency. https://www.epa.gov/newsreleases/epa-finalizes-affordable -clean-energy-rule-ensuring-reliable-diversified-energy. Accessed on August 28, 2019.

Epstein, Edward S. 1978. "Beginnings of the National Climate Program." *Bulletin of the American Meteorological Society* 59(11): 1402–1405.

Evarts, Holly. 2016. "Increasing Tornado Outbreaks: Is Climate Change Responsible?" State of the Planet. Columbia University. http://blogs.ei.columbia.edu/2016 /12/01/increasing-tornado-outbreaks-is-climate-change -responsible/. Accessed on September 6, 2019.

"Events and Conferences." 2019. Climate Diplomacy. https:// www.climate-diplomacy.org/events. Accessed on September 16, 2019."Explaining Extreme Events." [annual report]. American Meteorological Society. https://www.ametsoc .org/ams/index.cfm/publications/bulletin-of-the-american -meteorological-society-bams/explaining-extreme-events -from-a-climate-perspective/. Accessed on September 6, 2019. (Editors vary by issue.)

"Extended Reconstructed Sea Surface Temperature (ERSST) v4." 2018. National Centers for Environmental Education. https://www.ncdc.noaa.gov/data-access/marineocean-data /extended-reconstructed-sea-surface-temperature-ersst-v4. Accessed on September 6, 2019.

"Facts and Statistics: Wildfires." n.d. Insurance Information Institute. https://www.iii.org/fact-statistic/facts-statistics -wildfires. Accessed on September 6, 2019.

Field, Christopher B., and Vicente R. Barros, eds. 2014. "Climate Change 2014. Impacts, Adaptation, and Vulnerability. Part A: Global and Sectoral Aspects." Working Group II Contribution to the Fifth Assessment Report of the Intergovernmental Panel on Climate Change. New York: Cambridge University Press. https://www.ipcc .ch/site/assets/uploads/2018/02/WGIIAR5-PartA_FINAL .pdf. Accessed on September 5, 2019.

Fleshler, David. 2018. "The World Has Never Seen a Category 6 Hurricane. But the Day May Be Coming." *Los Angeles Times*. http://www.latimes.com/nation/la-na -hurricane-strenth-20180707-story.html. Accessed on September 6, 2019.

Fletcher, Luke, et al. 2018. "Beyond the Cycle." Carbon
Disclosure Project. https://6fefcbb86e61af1b2fc4-c70d8e
ad6ced550b4d987d7c03fcdd1d.ssl.cf3.rackcdn.com/cms
/reports/documents/000/003/858/original/CDP_Oil_and
_Gas_Executive_Summary_2018.pdf (Executive summary
only). Accessed on September 25, 2019.

"For John Ehrlichman." 1969. The White House. https://
capitalresearch.org/app/uploads/2012/11/tpr-1107-Daniel
-P-Moynihan-110520-article-memo-on-global-warming-to
-Nixon-690917.pdf. Accessed on August 28, 2019.

"Former Climate Change Deniers, What Changed Your
Mind?" 2017. Reddit. https://www.reddit.com/r/AskReddit
/comments/5zvuxx/former_climate_change_deniers_what
_changed_your/. Accessed on September 2, 2019.

Franz, Wendy E. 1998. "Science, Skeptics and Non-state
Actors in the Greenhouse." Global Environmental
Assessment Report. John F. Kennedy School of
Government. Harvard University. https://www.belfercenter
.org/sites/default/files/legacy/files/Science%20Skeptics
%20and%20Non-State%20Actors%20in%20the
%20Greenhouse%20-%20E-98-18.pdf. Accessed on
September 1, 2019.

Freitas, Vania, et al. 2010. "Temperature Tolerance and
Energetics: A Dynamic Energy Budget-Based Comparison
of North Atlantic Marine Species." *Philosophical
Transactions of the Royal Society B* 365: 3553–3565.
doi:10.1098/rstb.2010.0049. https://royalsocietypublishing
.org/doi/pdf/10.1098/rstb.2010.0049. Accessed on
September 14, 2019.

Gasparrini, Antonio, et al. 2017. "Projections of Temperature-
Related Excess Mortality under Climate Change
Scenarios." *Lancet Planetary Health* 19(9): e360–e367.
https://www.thelancet.com/action/showPdf?pii=S2542
-5196%2817%2930156-0. Accessed on September 9, 2019.

Germain, Tiffany, et al. 2013. "The Anti-Science Climate Denier Caucus." ThinkProgress. https://thinkprogress.org /the-anti-science-climate-denier-caucus-732ec3a2a4d4/. Accessed on August 31, 2019.

"Global Climate Coalition." n.d. Desmog. https://www .desmogblog.com/global-climate-coalition#s2. Accessed on September 1, 2019.

"Global Climate Coalition: Climate Denial Legacy Follows Corporations." 2019. Climate Investigations Center. https://climateinvestigations.org/global-climate-coalition -industry-climate-denial/. Accessed on September 1, 2019.

"Global Climate Report—May 2018." 2018. National Centers for Environmental Information. https://www.ncdc .noaa.gov/sotc/global/201805. Accessed on September 6, 2019.

"Global Temperature." 2019. Global Climate Change. Vital Signs of the Planet. NASA. https://climate.nasa.gov/vital -signs/global-temperature/. Accessed on September 6, 2019.

"Global Warming Disinformation Database." 2019. Desmog. https://www.desmogblog.com/global-warming-denier -database. Accessed on August 30, 2019.

"Global Warming of 1.5°C." 2016. Intergovernmental Panel on Climate Change. https://www.ipcc.ch/sr15/. Accessed on September 17, 2019.

Goodall, Chris. 2010. *How to Live a Low-Carbon Life: The Individual's Guide to Tackling Climate Change.* 2nd ed. London and Washington, DC: Earthscan.

Grant, Andrew, and Mike Coffin. 2019. "Breaking the Habit." Carbon Tracker. https://www.carbontracker.org/reports /breaking-the-habit/. Accessed on September 25, 2019.

Greenberg, Daniel S. [1969] 1999. *The Politics of Pure Science.* New York: New American Library; Chicago: University of Chicago Press.

Grimm, Nancy B., et al. 2013. "The Impacts of Climate Change on Ecosystem Structure and Function." *Frontiers in Ecology and the Environment* 11(9): 474–482. https://esajournals.onlinelibrary.wiley.com/doi/pdf/10.1890/120282. Accessed on September 14, 2019.

Groffman, Peter M., et al. 2014. "Ecosystems, Biodiversity, and Ecosystem Services." In *Climate Change Impacts in the United States: The Third National Climate Assessment*, edited by J. M. Melillo, et al., 195–219. U.S. Global Change Research Program. doi:10.7930/J0TD9V7H. http://nca2014.globalchange.gov/report/sectors/ecosystems. Accessed on September 14, 2019.

Gruber, Joachim, and Marianne Steenken. 2018. "Residence Time of Carbon Dioxide in the Atmosphere." Acamedia. http://www.acamedia.info/sciences/sciliterature/globalw/residence.htm. Accessed on September 3, 2019.

Guo, Yuming, et al. 2018. "Quantifying Excess Deaths Related to Heatwaves under Climate Change Scenarios: A Multicountry Time Series Modelling Study." *PLoS Medicine* 15(7): e1002629. https://journals.plos.org/plosmedicine/article?id=10.1371/journal.pmed.1002629#pmed-1002629-t001. Accessed on September 9, 2019.

Gutin, Ori, and Brendan Ingargiola. 2015. "Fact Sheet—Timeline of Progress Made in President Obama's Climate Action Plan." Environmental and Energy Study Institute. https://www.eesi.org/papers/view/fact-sheet-timeline-progress-of-president-obama-climate-action-plan. Accessed on August 29, 2019.

"Holding Major Fossil Fuel Companies Accountable for Nearly 40 Years of Climate Deception and Harm." n.d. Union of Concerned Scientists. https://www.ucsusa.org/resources/holding-major-fossil-fuel-companies-accountable. Accessed on December 11, 2019.

Houghton, J. T., et al., eds. 2001. "Climate Change 2001: The Scientific Basis." Intergovernmental Panel for Climate Change. New York: Cambridge University Press. https://www.ipcc.ch/site/assets/uploads/2018/03/WGI_TAR_full_report.pdf. Accessed on September 3, 2019.

"How Does the Medieval Warm Period Compare to Current Global Temperatures?" 2015. Skeptical Science. https://skepticalscience.com/medieval-warm-period.htm. Accessed on September 4, 2019.

"How to Talk to a Climate Skeptic: Responses to the Most Common Skeptical Arguments on Global Warming." 2019. Grist. https://grist.org/series/skeptics/. Accessed on August 30. 2019.

"In-depth: Experts Assess the Feasibility of 'Negative Emissions.'" 2016. CarbonBrief. https://www.carbonbrief.org/in-depth-experts-assess-the-feasibility-of-negative-emissions. Accessed on September 17, 2019.

"Information Council for the Environment." 1991. [no title]. https://www.ucsusa.org/sites/default/files/attach/2015/07/Climate-Deception-Dossier-5_ICE.pdf. Accessed on August 31, 2019.

"Information Council for the Environment." n.d. Desmog. https://www.desmogblog.com/information-council-environment. Accessed on August 31, 2019.

Jervey, Ben. 2015. "Fossil Fuel Industry's Global Climate Science Communications Plan in Action: Polluting the Classroom." Desmog. https://www.desmogblog.com/2015/02/27/api-global-climate-science-communications-plan-action-fossil-fuels-clasroom. Accessed on August 31, 2019.

Kennedy, Brian, and Meg Hefferon. 2019. "U.S. Concern about Climate Change Is Rising, but Mainly among Democrats." Pew Research Center. https://www.pewresearch.org/fact-tank/2019/08/28/u-s-concern-about

-climate-change-is-rising-but-mainly-among-democrats/. Accessed on August 28, 2019.

Kingsley, Patrick. 2018. "Trump Says California Can Learn from Finland on Fires. Is He Right?" *New York Times.* https://www.nytimes.com/2018/11/18/world/europe /finland-california-wildfires-trump-raking.html. Accessed on September 8, 2019.

Kirchmeier-Young, N., et al. 2019. "Attribution of the Influence of Human-Induced Climate Change on an Extreme Fire Season." *Earth's Future* 7(1): 2–10. https:// agupubs.onlinelibrary.wiley.com/doi/pdf/10.1029 /2018EF001050. Accessed on September 8, 2019.

Kossin, James P. 2018. "A Global Slowdown of Tropical-cyclone Translation Speed." *Nature* 558: 104–107.

Lewin, Bernie. 2010. "Timeline: Chronology of Climate Change Alarmism in Climate Science." Enthusiasm, Scepticism and Science. https://enthusiasmscepticismscience .wordpress.com/chronology-of-climate-change-science/. Accessed on August 28, 2019.

Magnason, Andri Snær. 2019. "The Glaciers of Iceland Seemed Eternal. Now a Country Mourns Their Loss." *The Guardian.* https://www.theguardian.com/commentisfree /2019/aug/14/glaciers-iceland-country-loss-plaque-climate -crisis. Accessed on September 5, 2019.

Malcolm, Jay R., and Louis F. Pitelka. 2000. "Ecosystems and Global Climate Change: A Review of Potential Impacts on U.S. Terrestrial Ecosystems and Biodiversity." Pew Center on Global Climate Change. https://www.c2es.org/site /assets/uploads/2001/12/env_ecosystems.pdf. Accessed on September 14, 2019.

Martinich, Jeremy, and Allison Crimmins. 2019. "Climate Damages and Adaptation Potential across Diverse Sectors of the United States." *Nature Climate Change*

9(5): 397–404. https://www.thecomingsfoundation.org
/resources/Hamful-Efects-of-CC-in-the-US.pdf. Accessed
on September 15, 2019.

McCarthy, James J., et al., eds. 2001. "Climate Change 2001:
Impacts, Adaptation, and Vulnerability." Intergovernmental
Panel on Climate Change. New York: Cambridge
University Press. https://www.ipcc.ch/site/assets/uploads
/2018/03/WGII_TAR_full_report-2.pdf. Accessed on
September 14, 2019.

McNeill, Aislinn. n.d. "Annual Disaster/Death Statistics for
US Storms." Texas Tech University. https://www.depts
.ttu.edu/nwi/research/DebrisImpact/Reports/DDS.pdf.
Accessed on September 13, 2019.

Moberg, Anders, et al. 2005. "Highly Variable Northern
Hemisphere Temperatures Reconstructed from Low-and
High-resolution Proxy Data." *Nature* 433(7026): 613–617.

Moon, Emily. 2019. "The Faulty Foundations of Trump's
Threats to Withhold FEMA Funding from Wildfire
Victims." Pacific Standard. https://psmag.com/news/the
-faulty-foundations-of-trumps-threats-to-withhold-fema
-funding-from-wildfire-victims. Accessed on September 6.
2019.

Mulvey, Kathy, and Seth Shulman. 2015. "The Climate
Deception Dossiers." The Indispensable Nation. http://
theindispensablenation.com/the-climate-deception-dossiers
.html. Accessed on August 31, 2019.

Nakićenović, Nebojša, et al. 2000. "Emissions Scenarios."
Intergovernmental Panel on Climate Change. New York:
Cambridge University Press. https://www.ipcc.ch/site/assets
/uploads/2018/03/emissions_scenarios-1.pdf. Accessed on
September 3, 2019.

National Academies of Sciences, Engineering, and Medicine.
2016. *Attribution of Extreme Weather Events in the Context
of Climate Change*. Washington, DC: National Academies

Press. doi:10.17226/21852. https://www.nap.edu /download/21852. Accessed on September 6, 2019.

National Research Council. 2015a. *Climate Intervention: Carbon Dioxide Removal and Reliable Sequestration.* Washington, DC: National Academies Press. doi:10.17226/18805.

National Research Council. 2015b. *Climate Intervention: Reflecting Sunlight to Cool Earth.* Washington, DC: National Academies Press. doi:10.17226/18988.

"1996 Global Climate Coalition: An Overview and Attached Reports." n.d. DocumentCloud. https://www.documentcloud .org/documents/5453339-1996-GCC-Overview-and-Reports .html. Accessed on September 1, 2019.

Nuccitelli, Dana. 2018. "New Study Finds Incredibly High Carbon Pollution Costs." Citizens Climate Lobby. https:// citizensclimatelobby.org/new-study-finds-incredibly-high -carbon-pollution-costs/. Accessed on September 15, 2019.

"Obama Administration Launches Quadrennial Energy Review." 2014. The White House. https:// obamawhitehouse.archives.gov/the-press-office/2014/01 /09/obama-administration-launches-quadrennial-energy -review. Accessed on August 29, 2019.

"Oil and Natural Gas Sector: Emission Standards for New, Reconstructed, and Modified Sources Review." 2019. Environmental Protection Agency. https://www.epa.gov /sites/production/files/2019-08/documents/frn_oil_and _gas_review_2060-at90_nprm_20190828revised_d.pdf. Accessed on August 30, 2019.

"Overview of the Clean Power Plan." 2015. Environmental Protection Agency. https://archive.epa.gov/epa /cleanpowerplan/fact-sheet-overview-clean-power-plan .html. Accessed on August 29, 2019.

"The Paris Agreement." 2019. United Nations Climate Change. https://unfccc.int/process-and-meetings/the

-paris-agreement/the-paris-agreement. Accessed on August 30, 2019.

Partain, James L., et al. 2016. "An Assessment of the Role of Anthropogenic Climate Change in the Alaska Fire Season of 2015." *Bulletin of the American Meteorological Society* 97(12): S14–S18. https://journals.ametsoc.org/doi/pdf/10 .1175/BAMS-D-16-0149.1. Accessed on September 8, 2019.

"Paying for Carbon Letter." 2015. DocumentCloud. http:// www.documentcloud.org/documents/2091463-paying-for -carbon-letter.html. Accessed on September 2, 2019.

"The Pigou Club." 2012. Notes on Liberty. https:// notesonliberty.com/2012/12/23/the-pigou-club/. Accessed on September 16, 2019.

Polansky, Anne. 2017. "ExxonMobil and Climate Change: A Story of Denial, Delay, and Delusion, Told in Forms 10-K (2009–2016)—Part Three (D): 2012." Government Accountability Project. https://www.whistleblower.org /politicization-of-climate-science/global-warming-denial -machine/exxonmobil-and-climate-change-a-story-of -denial-delay-and-delusion-told-in-forms-10-k-2009-2016 -part-three-d-2012/. Accessed on September 2, 2019.

Porter, John R., et al. 2014. "Food Security and Food Production Systems." In *Climate Change 2014: Impacts, Adaptation, and Vulnerability*, Part A: *Global and Sectoral Aspects*, Contribution of Working Group II to the Fifth Assessment Report of the Intergovernmental Panel on Climate Change, edited by C. B. Field et al., 485–533. New York: Cambridge University Press. https://www.ipcc .ch/site/assets/uploads/2018/02/WGIIAR5-Chap7_FINAL .pdf. Accessed on September 13, 2019.

"Predictions." 2006. Climate Change. http://www.ghgonline .org/predictions.htm. Accessed on September 3, 2019.

"Presidential Executive Order on Promoting Energy Independence and Economic Growth." 2017. The White

House. https://www.whitehouse.gov/presidential-actions
/presidential-executive-order-promoting-energy-independence
-economic-growth/. Accessed on August 29, 2019.

"The President's Climate Action Plan." 2013. The White
House. https://obamawhitehouse.archives.gov/sites/default
/files/image/president27sclimateactionplan.pdf. Accessed
on August 29, 2019.

"The President's Climate Data Initiative: Empowering
America's Communities to Prepare for the Effects of
Climate Change." 2014. The White House. https://
obamawhitehouse.archives.gov/the-press-office/2014/03/19
/fact-sheet-president-s-climate-data-initiative-empowering
-america-s-comm. Accessed on August 29, 2019.

"Projections for Carbon Dioxide." 2012. Association for
Canadian Educational Resources. http://www.acer-acre.ca
/resources/climate-change-in-context/introduction-2/global
/scientific-projections/projections-for-carbon-dioxide.
Accessed on September 3, 2019.

Regoli, Natalie. 2019. "20 Cap and Trade System Pros and
Cons." Vittana. https://vittana.org/20-cap-and-trade
-system-pros-and-cons. Accessed on September 16, 2019.

"Reports." 2019. Intergovernmental Panel on Climate
Change. http://www.ipcc.ch/reports/. Accessed on
September 2, 2019.

"Restoring the Quality of Our Environment. Report of
the Environmental Pollution Panel. President's Science
Advisory Committee." 1965. https://ozonedepletiontheory
.info/Papers/Revelle1965AtmosphericCarbonDioxide.pdf.
Accessed on August 28, 2019.

Revkin, Andrew C. 2009. "Industry Ignored Its Scientists
on Climate." *New York Times*. https://www.nytimes.com
/2009/04/24/science/earth/24deny.html. Accessed on
September 1, 2019.

Rich, Nathaniel. 2018. "Losing Earth: The Decade We Almost
Stopped Climate Change." *New York Times*. https://www

.nytimes.com/interactive/2018/08/01/magazine/climate
-change-losing-earth.html. Accessed on August 28, 2019.

Ricke, Katharine, et al. 2018. "Country-level Social Cost of
Carbon." *Nature Climate Change* 8(10): 895–900. Preprint
http://www.cobham-erc.eu/wp-content/uploads/2019/04
/preprint_Ricke2018_country_level_scc.pdf. Accessed on
September 15, 2019.

Santos, Rita. 2019. *Geoengineering: Counteracting Climate
Change.* New York: Greenhaven Publishing.

Savage, Karen. 2019. "Global Climate Coalition: Fighting
Global Climate Action in Favor of Fossil Fuels' Survival."
Climate Liability News. https://www.climateliabilitynews
.org/2019/04/25/gcc-global-climate-coalition-un-fossil
-fuels/. Accessed on September 1, 2019.

Schwartz, John. 2016. "Tillerson Led Exxon's Shift on
Climate Change; Some Say 'It Was All P.R.'" *New York
Times.* https://www.nytimes.com/2016/12/28/business
/energy-environment/rex-tillerson-secretary-of-state-exxon
.html. Accessed on September 2, 2019.

Seeley, Jacob T., and David M. Romps. 2015. "The Effect of
Global Warming on Severe Thunderstorms in the United
States." *Journal of Climate* 28(6): 2443–2458. https://
journals.ametsoc.org/doi/pdf/10.1175/JCLI-D-14-00382
.1. Accessed on July 17, 2018.

Shabecoff, Philip. 1988. "Global Warming Has Begun, Expert
Tells Senate." *New York Times.* https://www.nytimes.com
/1988/06/24/us/global-warming-has-begun-expert-tells
-senate.html. Accessed on August 28, 2019.

Sisson, Patrick, Megan Barber, and Alissa Walker. 2019. "101
Ways to Fight Climate Change." Curbed. https://www
.curbed.com/2017/6/7/15749900/how-to-stop-climate
-change-actions. Accessed on September 16, 2019.

Skuce, Andy. 2012. "Big Oil and the Demise of Crude
Climate Change Denial." Skeptical Science. https://

skepticalscience.com/bigoil.html. Accessed on September 2, 2019.

Smith, Kirk R., et al. 2014. "Human Health: Impacts, Adaptation, and Co-benefits." In *Climate Change 2014: Impacts, Adaptation, and Vulnerability*, Part A: *Global and Sectoral Aspects*, Contribution of Working Group II to the Fifth Assessment Report of the Intergovernmental Panel on Climate Change," edited by C. B. Field et al., 709–754. New York: Cambridge University Press.

Smith, Nick. 2014. "Humans (Surprise!) Biggest Cause of Glacier Loss." GlacierHub. https://glacierhub.org/2014 /08/28/humans-surprise-biggest-cause-of-glacier-loss/. Accessed on September 5, 2019.

Somanander, Tanya. 2016. "President Obama: The United States Formally Enters the Paris Agreement." The White House. https://obamawhitehouse.archives.gov/blog/2016 /09/03/president-obama-united-states-formally-enters-paris -agreement. Accessed on August 30, 2019.

Soon, Willie, and Sallie Baliunas. 2003. "Proxy Climatic and Environmental Changes of the Past 1000 Years." *Climate Research* 23(2): 89–110. https://www.int-res.com/articles /cr2003/23/c023p089.pdf. Accessed on August 31, 2019.

"Statement by President Trump on the Paris Climate Accord." 2017. The White House. https://www.whitehouse.gov /briefings-statements/statement-president-trump-paris -climate-accord/. Accessed on August 30, 2019.

"Statement of Dr. James Hansen, Director, NASA Goddard Institute for Space Sciences." 1988. http://image.guardian .co.uk/sys-files/Environment/documents/2008/06/23 /ClimateChangeHearing1988.pdf. Accessed on August 28, 2019.

Tabuchi, Hiroko. 2019. "Oil Giants, under Fire from Climate Activists and Investors, Mount a Defense." *New York Times*.

https://www.nytimes.com/2019/09/23/climate/oil-industry
-climate-investment.html. Accessed on September 25, 2019.

Tett, Simon F. B., et al. 2018. "Anthropogenic Forcings and
Associated Changes in Fire Risk in Western North America
and Australia during 2015/16." *Bulletin of the American
Meteorological Society* 99(1): S60–S64. https://journals
.ametsoc.org/doi/pdf/10.1175/BAMS-D-17-0096.1.
Accessed on September 8, 2019.

Tippett, Michael K., Chiara Lepore, and Joel E. Cohen.
2016. "More Tornadoes in the Most Extreme U.S. Tornado
Outbreaks." *Science*. doi: 10.1126/science.aah7393. http://
lab.rockefeller.edu/cohenje/assets/file/416TippettLeporeCo
henExtremeTornadoOutbreaks_SuppMatScience2016.pdf.
Accessed on September 6, 2019.

Tol, Richard S. J. 2014. *Climate Economics: Economic
Analysis of Climate, Climate Change and Climate Policy*.
Cheltenham, UK: Edward Elgar.

Trapp, Robert J., et al. 2007. "Changes in Severe
Thunderstorm Environment Frequency during the 21st
Century Caused by Anthropogenically Enhanced Global
Radiative Forcing." *PNAS* 104 (50): 19719–19723. http://
www.pnas.org/content/104/50/19719#sec-2. Accessed on
December 6, 2019.

Trtanj, Juli M., et al. 2016. "Climate Impacts on Water-
Related Illness." In *The Impacts of Climate Change on
Human Health in the United States: A Scientific Assessment*.
U.S. Global Change Research Program, 157–188.
doi:10.7930/J03F4MH. https://health2016.globalchange
.gov/water-related-illness. Accessed on September 13, 2019.

"2001 State Department Briefing for Global Climate
Coalition Meeting." n.d. DocumentCloud. https://www
.documentcloud.org/documents/4407192-Global-Climate
-Coalition-Meeting-2001.html#document/p3/a420404.
Accessed on September 1, 2019.

"U.S. DOT and EPA Propose Fuel Economy Standards for MY 2021–2026 Vehicles." 2018. U.S. Department of Transportation. https://www.transportation.gov/briefing -room/dot4818. Accessed on August 29, 2019.

"US 116th Congress Sets New Record for Members with STEM Backgrounds." Cambridge Core. https://www .cambridge.org/core/journals/mrs-bulletin/article/us -116th-congress-sets-new-record-for-members-with-stem -backgrounds/6BAADCDA3CAB1925EEA62FDACF24F 7C4/core-reader. Accessed on August 28, 2019.

Walker, Joe. 1998. "Global Climate Science Communications." https://insideclimatenews.org/sites /default/files/documents/Global%20Climate%20Science %20Communications%20Plan%20%281998%29.pdf. Accessed on August 31, 2019.

Wayne, Graham. 2013. "The Beginner's Guide to Representative Concentration Pathways." Skeptical Science. https://skepticalscience.com/docs/RCP_Guide.pdf. Accessed on September 4, 2019.

Weart, Spencer. 2019. "Government: The View from Washington, DC." The Discovery of Global Warming. American Institute of Physics. https://history.aip.org /climate/Govt.htm#L_M046. Accessed on August 28, 2019.

Westerling, A. L., et al. 2006. "Warming and Earlier Spring Increase Western U.S. Forest Wildfire Activity." *Science* 313(5789): 940–943. doi:10.1126/science.1128834. https://science.sciencemag.org/content/sci/313/5789/940 .full.pdf. Accessed on September 8, 2019.

"What Do We Know about Wildfire Attribution and Climate Change?" 2018. Energy Innovation. https:// energyinnovation.org/2018/09/25/what-do-we-know -about-wildfire-attribution-and-climate-change/. Accessed on September 8, 2019.

"Where Carbon Is Taxed." 2018. Carbon Tax Center. https:// www.carbontax.org/where-carbon-is-taxed/#Other. Accessed on September 17, 2019.

"Where Is Earth's Water?" n.d. U.S. Geological Survey. https://www.usgs.gov/special-topic/water-science-school /science/where-earths-water. Accessed on September 5, 2019.

Williams, A. Park, et al. 2019. *Earth's Future*. doi:10.1029/2019EF001210. https://agupubs.onlinelibrary .wiley.com/doi/pdf/10.1029/2019EF001210. Accessed on September 8, 2019.

"Willie Soon." 2019. Desmog. https://www.desmogblog.com /willie-soon. Accessed on August 31, 2019.

World Bank, Ecofys, and Vivid Economics. 2017. "State and Trends of Carbon Pricing 2017." Washington, DC: World Bank. doi:10.1596/978-1-4648-1218-7. https:// openknowledge.worldbank.org/bitstream/handle/10986 /28510/wb_report_171027.pdf. Accessed on September 17, 2019.

Ye, Jason. 2013. "Options and Considerations for a Federal Carbon Tax." Center for Climate and Energy Solutions. https://www.c2es.org/document/options-and -considerations-for-a-federal-carbon-tax/. Accessed on September 17, 2019.

Yoon, Jin-Ho, et al. 2015. "Extreme Fire Season in California: A Glimpse into the Future?" *Bulletin of the American Meteorological Society* 96(12): S5–S9.

Zhou, Tian Jun, et al. 2011. "A Comparison of the Medieval Warm Period, Little Ice Age and 20th Century Warming Simulated by the FGOALS Climate System Model." *Chinese Science Bulletin* 56(28–29): 3028–3041. doi:10.1007/s11434-011-4641-6. https://core.ac.uk /download/pdf/81724026.pdf. Accessed on September 4, 2019.

Introduction

Climate change is a topic that evokes strong feelings in many individuals. Those feelings range from considerable concern to rejection. This chapter provides essays from individuals who have been involved in some aspect of the climate change debate or engaged in some specific aspect of that issue.

Cemetery Management Today or Reinternment Management Tomorrow
Jennifer Blanks

Cemeteries are time stamps and places for reflection that connect us to our loved ones. However, ecological imbalances could limit access to these places. As climate change progresses, rising sea levels, flooding, and hurricanes will increase in frequency, causing significant damage to vulnerable cemeteries in flood-prone, coastal communities. Strategic mitigation planning for these vulnerable cemeteries is required.

Flooding Impacts

Hurricane Katrina revealed the catastrophic impact of severe flooding on cemeteries. In August 2005, Hurricane Katrina

---

Protesters carry placards and banners at the Climate Strike Rally and March in downtown San Francisco, California on September 20, 2019. (Andrei Gabriel Stanescu/Dreamstime.com)

leveled Coastal Mississippi and Louisiana. In the days following landfall, haunting images of the damage surfaced and shocked the world. Hurricane Katrina impacted roughly 1,500 gravesites (Lovekamp, Foster, and Di Naso 2016). Floodwaters dislodged coffins from their tombs, causing many to reopen (Chen 2008). Following massive flooding, cemetery managers had the difficult task of identifying the deceased and reinterning them to the correct gravesite (Lovekamp, Foster, and Di Naso 2016). Documentation of coffin ownership and burial records were destroyed, making cemetery recovery even more challenging. Restoring caskets and tomb markers is costly for cemetery operators and uninsured families of the deceased.

In August 2016, Southern Louisiana experienced a series of thunderstorms. Shortly after floodwaters rose, coffins across several Louisiana parishes began rising as well. The amount of floodwater accumulated during the 2016 deadly storms displaced 300 tombs weighing 2,000 pounds each (Lau 2016).

The most vulnerable cemeteries are in rural, low-lying flood plains in unincorporated areas with poor infrastructure (Disaster Planning 2008). These same rural areas have small migratory populations with low visitation rates and have often been annexed (Disaster Planning 2008). In this case, gravesites are largely unmaintained until the next funeral service or abandoned altogether. These cemeteries usually have no visible boundaries, overgrown vegetation, poorly maintained or absent tomb markers. Lastly, rural cemeteries located in floodways suffer from mud deposits and shifted headstones (Disaster Planning 2008). Cemetery managers need maintenance plans and mitigation strategies that address likely damage areas after disasters.

Current Management and Practices

The agencies who manage cemeteries are the state government, National Park Services, and Federal Emergency Management Agency (FEMA). Currently, state government manages Louisiana's cemeteries under the Cemetery Care Act of 1974. The

Cemetery Care Act requires cemetery operators to incorporate the graveyard and gravesite with a state-issued license (Maintenance Funds 2019). Operators must provide routine gravesite documentation to receive state care funds. States also regulate the location of cemeteries as well. Allowing the state to control cemetery placement is significant because the state has the authority to place publicly owned cemeteries in highly vulnerable areas (Disaster Planning 2008). In regards to disaster management, FEMA collects photographic documentation following natural disasters. FEMA also has an Environmental Planning and Historic Preservation team who help mitigate, prepare, and restore cemeteries post-disaster ("Historic City Cemeteries Can Once Again Rest in Peace" 2011). The team also helps manage human reinternment following disasters as seen in the Louisiana floods ("FEMA Reinternment Assistance" 2016). FEMA introduces cemetery operators to an abundance of resources and tools that allow them to maintain cemeteries before and after severe flooding and other disasters.

The combination of GIS technology and spatial data dramatically transformed cemetery management (Stein 2006). Prior to GIS implementation in cemetery management, managers struggled to maintain records of their property, tomb, and marker maintenance, and burial location records. Today, GIS allows managers to input identifying characteristics such as plot markers, death certificates, and coordinates on a map. In regards to flooding and disaster planning with coffin displacement, GIS greatly aids during the reinternment process and overall recovery.

Limitations and Policy Recommendations

A limitation is the cost of integrating GIS technology into cemetery management. GIS software, high-quality cameras, and GPS tools can be quite expensive. Cemeteries in geographically and socially vulnerable areas will have a difficult time finding this form of maintenance if the state does not provide enough funding. The process of collecting and maintaining data can be tedious and time-intensive. Cemetery operators need to allocate time for training to make the most of the technology. The state will not approve recovery funds for cemeteries without state-issued licenses. As a result, an abandonment occurs when the publicly owned cemetery is no longer maintained and evidence of neglect may soon emerge (*Adams v. State* 1957).

Appropriate policy recommendations to improve cemetery management as climate change continues to be a threat address funding allocation, training, and technical assistance needs. State and federal government entities should designate funding for local cemetery managers as well as training and technical assistance. These assistance programs can be widely adapted to other cemeteries throughout the United States and should include guidance on creating mitigation plans. Cemetery mitigation planning prior to a disaster can minimize damage or at the least make recovery manageable. Lastly, research and community-based organizations require more funding to create and apply advanced technology that will transform cemetery management.

References

Adams v. State. 1957. 95 Ga. App. 295 (1957).

Chen, Stephanie. 2008. "In Louisiana, the Search Goes on for Lost Coffins." *Wall Street Journal.* https://www.wsj.com /articles/SB122005767852185305. Accessed on September 26, 2019.

"Disaster Planning." 2008. Chicora Foundation. https:// www.chicora.org/disaster-planning.html. Accessed on September 26, 2019.

"FEMA Reinternment Assistance." 2016. FEMA. https:// www.fema.gov/news-release/2016/09/14/fema-reinterment -assistance. Accessed on September 26, 2019.

"Historic City Cemeteries Can Once Again Rest in Peace."
2011. FEMA. https://www.fema.gov/news-release/2011
/06/14/historic-city-cemeteries-can-once-again-rest-peace.
Accessed on September 26, 2019.

Lau, Maya. 2016. "After Flood Uproots Nearly 300 Graves,
the Cost of Repair May Fall on Small Churches, Family."
The Advocate. https://www.theadvocate.com/baton_rouge
/article_3635288c-73a7-11e6-9fda-73a1304e462e.html.
Accessed on September 26, 2019.

Lovekamp, William, Gary Foster, and Steven Di Naso. 2016.
"Preserving the Dead: Cemetery Preservation and Disaster
Planning." Natural Hazards Workshop, Natural Hazards
Center, Broomfield, CO. https://hazards.colorado.edu
/article/preserving-the-dead-cemetery-preservation-and
-disaster-planning. Accessed on September 26, 2019.

"Maintenance Funds." 2019. "Cemetery Maintenance."
US Legal. https://cemeteries.uslegal.com/cemetery
-maintenance/maintenance-funds/. Accessed on September
26, 2019.

Stein, James. 2006. "GIS Tools for Cemetery Management."
National Park Service. ESRI International Users
Conference, NPS Cultural Resources GIS. https://www
.nps.gov/crgis/proj_alexandria_cemetery.pdf. Accessed on
September 26, 2019.

*Jennifer Blanks is a proud native of New Orleans, Louisiana. She
is a doctoral student studying urban development and regional science at Texas A&M University. Her research interests are historical
preservation of cemeteries in African American communities and
the impacts natural disasters have on the graves using GIS and
remote sensing. Through her research, Blanks wishes to encourage
women, especially women of color, to learn and apply geospatial
technology to increase their representation as geographic information system users across all academic and professional disciplines.*

Comatose in Climate Catastrophe
Olivia Cooper

We hear calls-to-action about climate change at different volumes based more on their perceived distance rather than actual threat level. Catastrophic yet temporally and spatially distant threats like climate change are whispers compared to minor yet immediate threats such as losing status or respect that we face when consuming, self-silencing, or otherwise protecting ourselves at Earth's expense. Threats such as losing social status or the respect of our peers, while they seem insignificant compared to the life-threatening effects of climate change, are close to us in space and time and are more along the lines of threats our brains are built to respond to at this phase in our evolution (Griskevicius and Kenrick 2013, 372–386). Effects tend to have a lag in time of multiple decades and are often somewhat invisible as they involve long-term changes such as global temperature and sea-level rise.

As of a 2018 report, 70 percent of Americans who responded to the Yale Project on Climate Change Communication survey stated that they believe global warming is happening, and 57 percent agreed that it is mostly caused by human activities (Marlon et al. 2018). However, many people fail to speak up about climate change for fear of alienating themselves from their peers, as they inaccurately assume their opinions on climate change are in the minority, and expect and are afraid to appear incompetent, alarmist, or ignorant. One experiment found that the threat of losing respect of their group because they perceive themselves to hold a minority opinion caused participants to self-silence rather than participate in discussions on climate change (Geiger and Swim 2016). This self-silencing is motivated by humans' tendency to follow social norms and align themselves with the majority, which, at least in part, rises from fundamental evolutionary motives to make friends and attain or maintain status.

This perceived distance from the consequences of climate change has created a stagnancy or disengagement not necessarily in peoples' opinions on climate change, but in peoples' actions against climate change (Pahl et al. 2014). While acting against climate change may seem well-motivated by the fundamental drive to evade physical harm, its psychological distance allows other motivations to be activated, which take priority in determining decisions and behavior. For example, while purchasing a luxury, gas-guzzling car contradicts the motivations to evade physical harm and avoiding disease (by exacerbating the consequences of climate change, which in a variety of ways negatively impacts human health and harms us physically) as well as caring for family (by lessening future financial security in the long term), if motivations such as attaining status or acquiring a mate are activated, they may take priority and influence one's consumer decisions.

While some groups that are more proximal to climate change effects, such as Miami residents, may notice significant changes in sea level from year to year as more and more disastrous consequences are happening, they will not necessarily associate these changes with the causes of climate change, such as driving their cars or buying single-use plastics. Actions that further climate change are also blameless, as every consumer is partially responsible, and there is no one figurehead to collectively place blame on. Those who view themselves and their communities to be distant from the effects of climate change, tend to perceive the ways in which their behavior is furthering climate change as either not having a direct effect or having an effect on a distant group or place in space or time. This likely encompasses nearly everyone in some way but notably includes those in a position of privilege who can afford to ignore the consequences of climate change, such as members of WEIRD (Western, educated, industrialized, rich, democratic) groups. Bringing awareness to the psychological factors that impact our behavior with respect to climate change can help people self-reflect and may encourage self-change and repair responses.

However, simply decreasing psychological distance in all realms is not entirely effective in motivating action against climate change. While there is evidence that increased psychological distance from climate change increases the intensity of our emotional responses (McDonald, Chai, and Newell 2015), the relationship between perceived distance and climate change action is not so cut-and-dried. One study showed that participants were more likely to make self-change and repair actions with respect to mitigating climate change when they experienced self-conscious emotions such as shame or guilt, rather than basic emotions such as fear or anger (Ejelöv et al. 2018, 1–8). Additionally, the researchers conducted a 2 × 2 study comparing abstract versus concrete and proximal versus distal climate change communication methods, and found that describing a distal climate change consequence in a concrete manner was more effective at inspiring self-change and repair actions than describing a proximal climate change consequence in a concrete manner (Ejelöv et al. 2018). Our psychological and emotional responses to various climate change communication strategies are complex and although it is a significant factor, the way we respond does not correlate linearly with psychological distance.

The motivation to make friends or maintain social status is a fundamental evolutionary motive and more often than not is much less psychologically distant than the threat of climate change. Consistently, we respond to minor, immediate threats rather than the catastrophic, global threat of climate change at the expense of the Earth and our own futures. Climate change's psychological distance from us is an important aspect to consider for climate change communicators, whether they are our professors, politicians, scientists, artists, or others. Climate communicators and psychologists should aim to tap into self-change and repair emotions by triggering self-conscious emotional responses in their audiences and to shed some light on some of the complex ways that humans respond to and interact with climate change.

References

Ejelöv, E., et al. 2018. "Regulating Emotional Responses to Climate Change—A Construal Level Perspective." *Frontiers in Psychology* 9: 629. doi:10.3389/fpsyg.2018.00629.

Geiger, N., and J. K. Swim. 2016. "Climate of Silence: Pluralistic Ignorance as a Barrier to Climate Change Discussion." *Journal of Environmental Psychology* 47: 79–90. doi:10.1016/j.jenvp.2016.05.002.

Griskevicius, V., and D. T. Kenrick. 2013. "Fundamental Motives: How Evolutionary Needs Influence Consumer Behavior." *Journal of Consumer Psychology* 23(3): 372–386. doi:10.1016/j.jcps.2013.03.003.

Marlon, J., et al. 2018. "Yale Climate Opinion Maps 2018." YPCCC. https://climatecommunication.yale.edu/visualizations-data/ycom-us-2018/. Accessed on August 31, 2019.

McDonald, R. I., H. Y. Chai, and B. R. Newell. 2015. "Personal Experience and the 'Psychological Distance' of Climate Change: An Integrative Review." *Journal of Environmental Psychology* 44: 109–118. doi:10.1016/j.jenvp.2015.10.003.

Pahl, S., et al. 2014. "Perceptions of Time in Relation to Climate Change." *Wiley Interdisciplinary Reviews: Climate Change* 5(3): 375–388. doi:10.1002/wcc.272.

Olivia Cooper is a rising senior at Smith College majoring in astronomy and physics and concentrating in climate change. After earning her PhD in Astronomy, she plans to pursue a career in extragalactic astrophysics, while honing her skills as a science communicator.

Coastal Community Faces Effects of Climate Change, Sea-Level Rise
ChrisAnn Silver Esformes

It becomes difficult to deny our climate is changing when the effects of a warming atmosphere are creeping steadily toward your back door. I have spent nearly 42 years living in Manatee County—a community on the Gulf Coast of Florida in a region deemed one of the most vulnerable to sea-level rise in the United States by the National Oceanic and Atmospheric Administration. In recent years, I have regularly watched high tides swell over seawalls on sunny days—tides that wouldn't have touched the seawall 20 years prior. I have hunkered down during storms that dropped 10 inches of rain in a matter of hours. In addition to being a lifelong resident of Manatee County, I am a reporter on Anna Maria Island, a narrow, nine-mile long strip of land on the west end of our county, situated between Tampa Bay and the Intracoastal Waterway.

The island is divided into three municipalities—Anna Maria on the north end, Holmes Beach, the largest city, located in the middle of the island, and Bradenton Beach on the southern portion. Anna Maria was the first to be established, in 1923. Holmes Beach was incorporated in 1950 and Bradenton Beach in 1952. While the three small cities each have their own distinct character and separate municipal governments, they face similar issues, many of which I have listened to during meetings and reported on for the local paper. Bradenton Beach and Holmes Beach, the two cities that have been my news beat for the past three years, share a city engineer, Lynn Burnett, whose job is to develop systems of stormwater infiltration and to deal with flooding issues. Some might say this is a Herculean task given the changing environmental—and developmental—conditions of the island.

Burnett and her team deal with infrastructure mostly placed in the 1950s, on a barrier-island that is barely above sea level. The island was slowly developed on top of other failed development, creating a foundation of sand, asphalt, concrete, and other debris through which pipes and infiltration systems have been constructed to deal with the constant overflow of storm- and tidewater. The aging infrastructure is a constant source of concern for the cities, which must comply with Federal Emergency Management Agency (FEMA) regulations in order to be eligible for much-needed grants to build new infiltration systems that meet Florida Department of Environmental Protection standards. New construction also must be compliant with FEMA regulations so as not to risk lowered rankings with the National Flood Insurance Program (NFIP). And as the county and state promote the island as an undeveloped paradise, more people—including those who are looking through a lens of financial interest versus an environmental one—are buying million-dollar waterfront lots to build the biggest structures permitted.

FEMA land-use standards must be maintained for the cities to participate in FEMA's NFIP, which provides a discount on flood insurance premiums for the city and its property owners through a community rating system that evaluates flood-resistance on a citywide basis. Perhaps it is needless to say, as waters rise, so does the cost of flood insurance. And since the entire island is in a flood zone, flood insurance is required for anyone who owns a home on Anna Maria Island and also carries a mortgage.

Manatee County is considered the county most susceptible to sea-level rise in the United States, according to the study by the consulting group 427, which assesses the economic risk of climate change on business, financial, and government institutions, worldwide. The Center for Climate Integrity, an initiative by the Institute for Governance & Sustainable Development, estimates about $2 billion in seawall building or replacement will be required of private owners and local governments in Manatee County by 2040. In 2015, the state enacted legislation requiring local governments with a coastal element in their comprehensive plan to adopt criteria addressing sea-level rise. According to state statute, the peril-of-flood component must include engineering solutions that reduce flood risk in coastal areas, encourage the removal of vulnerable coastal property, such as mobile homes, from flood zones, and incorporate site development techniques to reduce flood losses.

There are two historic mobile home parks in Bradenton Beach that face future challenges as a result of changing legislation and rising flood waters. At a Manatee County Council of Governments meeting in January 2019, Sean Sullivan, executive director of the Tampa Bay Regional Planning Council, said an economic impact study conducted in 2016 by the TBRCP indicated that if Manatee County does nothing to address sea-level rise, by the year 2060, there could be $400 billion worth of property at risk. According to NOAA projections, also by 2060 the sea level will rise over existing seawalls in Bradenton Beach twice a day during high tide. The council has created a Peril of Flood project team, which researches sea-level rise and how to combat it as a unified front in the Tampa Bay region.

Counties and cities in the southwest Florida region have realized the urgency to team up to do what they can to mitigate the coming damage. Meanwhile, residents of the three island cities in Manatee County continue to note the changes that seem to be happening at an increasingly rapid rate and worry about how this will affect their island homes and businesses. At a May 2019 meeting of Anna Maria Island officials, Burnett said sea-level tracking technology has improved, and data shows the level is rising significantly faster than first projected. "We need to do something sooner, rather than later," she said. "There comes a tipping point, where if we wait we've waited too long."

At a city commission meeting that same month, Holmes Beach Mayor Judy Titsworth, whose family established Holmes Beach, said, "It doesn't really matter whether or not we believe in sea level rise—our kids and grandkids will, because they are going to be the ones dealing with its effects."

ChrisAnn Silver Esformes is a news reporter for a weekly newspaper, The Islander, *on Anna Maria Island, Florida. She received her MA in Mass Communication from the University of Florida. She is a native Floridian.*

A Call for Climate Literacy
Gregory Foy and Leigh Foy

Can you think of another time in history when science education was such an important stakeholder in the future of the planet? Can you think of another time in history when scientists have amassed such a significant amount of data indicating what is happening at a global scale and yet that science still faces extreme opposition within significant and powerful portions of the population? At this point, you might be thinking, "I thought that there was scientific uncertainty surrounding the science of climate change," and if you have been getting your information from certain sources, you might think that there still is controversy. But despite some very well organized, and well-funded disinformation campaign's claims, the level of scientific certainty has reached five sigma (Santer et al. 2019). This is what scientists frequently refer to as the "gold standard." This standard literally means that there is a 1 in 1,000,000 chance that humans are not responsible for the increase in CO_2 in our atmosphere. The scientific data on climate change is clear and not at all controversial! Not only has the scientific data reached such a level of certainty, but the public is beginning to react and demand a response.

Climate change, global warming, climate science, extreme weather—you have heard most of these terms used in social media, news media, movies, podcasts, and more, but if you don't know how intimately connected they are, you cannot completely understand the effects. This is the reason that climate science literacy is so important in society today because without this understanding people cannot make informed decisions. An incomplete understanding of climate science may be one of the most significant reasons that these terms have become so polarizing—many people aren't getting the scientific information (Mitchell et al. 2016). In fact, a consortium of newspapers in Florida has recently banded together to cover climate change stories in an effort to increase understanding of the effects of climate change on the residents of this state that will feel the impacts in significant ways (Harris 2019).

The science tells us that this generation of students will be the most impacted by climate change. According to the Fourth National Climate Assessment Report, "Decisions that decrease or increase emissions over the next few decades will set into motion the degree of impacts that will likely last throughout the rest of this century, with some impacts (such as sea level rise) lasting for thousands of years or even longer" (USGCRP 2018, 1360). This is why we so desperately need climate science education now to give our students the tools they need to tackle and cope with this significant environmental issue. The Next Generation Science Standards (NGSS) were established in 2013 by states and the top scientific organizations and uses the latest science education research to prepare students for STEM (science, technology, engineering, and math) careers in the 21st century. Climate change is a "core idea" in both middle school and high school standards. The National Science Teacher Association is also solidly behind climate change education in our schools. The National Center for Science Education and Penn State University recently published a study ("Mixed Messages: How Climate Change Is Taught in America's Public Schools" 2016) that shows climate change is not being taught correctly, if at all, while a recent NPR poll indicates that 80 percent of parents want their children to be taught about climate change (Kamenetz 2019).

The impacts of climate change are being felt around the globe, with the increased intensity of storms taking center stage. Hurricanes and typhoons stronger than ever before, record temperatures around the globe, long and strong heat waves, increases in number of fires and in the length of wildfire seasons, and significant droughts. Increases in ocean temperatures fuel bigger storms, and a decrease in the temperature differential between the Earth's poles and equator causes changes in the jet stream. These changes in the jet stream lead to extreme temperature variations that exacerbate droughts and floods, and cause heat waves and bitter cold days. The changes in climate have been described this way in the scientific community—it is getting hotter and colder, and drier and wetter.

Worldwide there are increasing numbers of climate marches, with one of the largest global climate strikes in September 2019 organized by youth. With all of these demands for action, it would seem obvious that climate science would be a significant part of the high school curriculum, but it is not. Studies show that teachers are not teaching climate science, since it does not fit neatly into a specific discipline, so no one is taking responsibility. Teachers shy away from covering controversial topics even though the science has achieved the aforementioned "gold standard." However, one of the biggest reasons for not teaching climate science is that teachers are not confident about the science, having never had formal training in the topic.

This is where one of the biggest problems lies but also where the solution is found: teacher education. We have put together a team of two college chemistry professors, a high school science teacher, and a professional writer to develop teacher workshops and learning materials ready for teacher use. A sample of the available materials is located on our website climateliteracyacademy.org. The materials located on this website are ready to use in the classroom to help explain things like ocean acidification, effects of warming oceans, and rising sea levels. These activities are just a sampling of what is available through teacher workshops.

If this climate change challenge seems completely over-whelming, we need only visit recent history to find examples of global cooperation in solving significant environmental issues. One of the world's greatest environmental success stories is the ongoing healing of the ozone layer through an international treaty called the Montreal Protocol. Just as world leaders banned the production of ozone-depleting chemicals, we must limit carbon-dioxide pollution from burning fossil fuels. But the first step in this process is education.

Our world is in danger, and it will take the elevation of climate literacy of the citizens of this planet if we are going to solve the greatest environmental challenge that humans have yet experienced.

References

Harris, Alex. 2019. "Florida's Leading News Organizations Announce Partnership for Covering Climate Change." https://www.sun-sentinel.com/news/florida/fl-ne -florida-climate-reporting-partnership-20190625 -dxn54v76qzhyvhttkwnai4jjwy-story.html. Accessed on September 28, 2019.

Kamenetz, Anya. 2019. "Most Teachers Don't Teach Climate Change; 4 in 5 Parents Wish They Did." NPR. https:// www.npr.org/2019/04/22/714262267/most-teachers -dont-teach-climate-change-4-in-5-parents-wish-they-did. Accessed on September 28, 2019.

Mitchell, Amy, et al. 2016. "Pathways to News." Pew Research Center. https://www.journalism.org/2016/07/07 /pathways-to-news/. Accessed on September 28, 2019.

"Mixed Messages: How Climate Change Is Taught in America's Public Schools." 2016. https://ncse.ngo/library -resource/mixed-messages-how-climate-change-is-taught -americas-public. Accessed on September 30, 2019.

Santer, B. D., et al. 2019. "Evidence of Human-Caused Climate Crisis Has Now Reached 'Gold Standard'-Level Certainty, Scientists Say." *Nature Climate Change* 9: 180–182.

USGCRP. 2018. *Fourth National Climate Assessment.* Vol. II. Washington, DC: U.S. Global Change Research Program.

Gregory Foy is an associate professor of chemistry at York College of Pennsylvania. He recently received the 2019 American Chemical Society "ACS-CEI Incorporation of Sustainability into Chemistry Education" award. Leigh Foy is a York Suburban High School science teacher and an adjunct professor at YCP.

Scientific Skepticism of Climate Change Models
Joel Grossman

Scientific skepticism is an antidote to dogmas and incontrovertible truths otherwise off-limits to discussion, reconsideration, or modification, believes President Obama's former climate advisor, 1973 physics Noble Prize winner Ivar Giaever. Giaever told the UK *Telegraph* newspaper: "Nothing is incontrovertible in science." In 2011 when the American Physical Society labeled global warming an incontrovertible truth, Giaever resigned in protest, saying global warming was wrongly transmuted into set-in-stone dogma (Sherwell 2011).

"Global warming has really become a new religion, because you can't discuss it," Giaever told the 2015 Lindau Nobel Laureate Meeting. Between 1898 and 1998 atmospheric carbon dioxide rose 72 ppm (24 percent) to 367 ppm, and Earth's average surface temperature increased 0.8 degrees (0.3 percent) from 288 to 288.8 Kelvin. "The temperature has been amazingly stable. In the same time period the number of people has increased in the world from 1.5 billion to over 7 billion. Is it possible that all the paved roads and cut-down forests have had an effect on the climate?" If 15 years of rising carbon dioxide (up 9.8 percent to 403 ppm) and stable temperature since 1998 were a scientific experiment, the carbon-dioxide (CO_2) climate change hypothesis would be rejected (Sherwell 2011).

In 1990, American physicist Richard Lindzen noted:

> an unusual degree of extremism associated with this issue. While environmental scares are not unheard of, few have been accompanied by recommendations that skepticism be stifled. . . . Certainly, we are dealing with significant changes in CO_2, but this alone need not be serious. CO_2 is a minor atmospheric constituent (about 0.03%), and as such, its variations might not be notably important. . . . Given the data alone, we would have little basis for alarm. The alarm arises instead from theoretical considerations—namely the so-called "greenhouse effect." (Lindzen 1990, 292)

The combined effect of ozone, methane, NO_2, and chlorofluorocarbons is "comparable to CO_2," but the "most important infrared absorbers are water vapor and liquid water in the form of cloud droplets" (Lindzen 1990). Clouds, which cover roughly half the Earth's surface, vex climate modelers because of contradictory greenhouse effects potentially of much higher magnitude (10×) than CO_2.

"Politicians and environmental scientists, especially in Europe, seem to have the mistaken idea that merely stabilizing the abundance of CO_2 in the air will stabilize the climate," wrote James Lovelock, inventor of instruments measuring trace levels of atmospheric gases, without which there would be neither debate nor treaties.

> In the real world the principal greenhouse gas that keeps us warm is not CO_2 but water vapor. Unlike what we call a permanent gas, the abundance of water vapor varies with temperature. . . . So intricate are the connections between the factors affecting climate change that we need to stay skeptical about the projections of climate models. . . .

What we were skeptical about was treating the mathematical model projections of climate change as if they were statistically significant . . . [and] the way politicians, and often their civil servants, took the IPCC projections almost as if written in stone like the message Moses brought down from the mountain. (Lovelock 2014)

"All models of the Earth are unfinished works still in construction," says Lovelock, whose Gaia hypothesis postulates a living planet with self-healing feedback loops. Major missing ingredients in climate models include: (1) the ocean, which "stores at least a thousand times more heat than the surface and atmosphere"; (2) living organisms processing and releasing gases; (3) carbon and oxygen chemistry. According to researchers at the George C. Marshall Institute: The drop in temperature between 1940 and 1970 casts doubt on whether greenhouse gases are always the main drivers of climate change. "A temperature increase of 0.5 C could be produced by an increase of 0.3 to 0.5 percent in the sun's brightness." "Natural variability" from unknown causes is even more important (Jastrow, Nierenberg, and Seitz 1990).

"So what about these giant climate models whose arithmetic is so difficult that only supercomputers can handle it?" asks Lovelock.

They include many coupled non-linear differential equations, and worse—non-linear partial differential equations. The very able mathematician Edward Lorenz showed in 1963 that dynamic systems containing more than three of these equations were prone to chaotic solutions. This means that starting from very similar initial conditions very different outcomes can occur, so tiny errors can be magnified in the course of calculation; you may have heard of the "butterfly effect." He showed that this is why models are not good at predicting the weather more than about a week ahead. . . . In climate science Lorenz found the same limitation for meteorology, and so did Robert May for ecosystems with more than two species present. (Lovelock 2014)

A physicist steeped in precise measurement, Giaever told Lindau's audience that 1880s thermometer baseline data was not accurate to tenths of a degree, the scale in the famous "hockey stick" diagram. European measurement mania, driven by D. G. Fahrenheit's 1714 mercury thermometer invention, has never had a solid methodological foundation for calculating global averages. The bias is toward "urban heat island" effects and warmth. The warmer Northern Hemisphere has 3,436 thermometers; versus 412 measuring a Southern Hemisphere with cooling trends (e.g., mid-latitudes of west-coast South America). Only 8 thermometers monitor the South Pole's falling temperatures and expanding ice sheets; 167 monitor the warmer, headline-grabbing North Pole (Giaever 2015).

"Never forget that the average temperature of the Earth is calculated from all of its climates, ranging from the icy cold of the polar regions to the hottest deserts—from about -90°C to 60°C, and that the average is dominated by the 70% of the Earth's surface that is ocean," writes Lovelock, who questions if global averages have local relevance, as Singapore is double (12°C) above the climate doomsday temperature and prospering. "For climate scientists to talk of consensus about their projections of climate change when they vary from 2 to 6 degrees Celsius of warming tells either of naivety or of irresistible pressures from national or other vested interests." Carbon-free nuclear power is stymied by overblown radiation fears, but European renewable energy entrepreneurs "see vast fortunes to be gathered from the subsidies offered by government" (Lovelock 2014).

"Save-the-planet," fueled by apocalyptic visions of climatic doom and true-believer zeal reminiscent of medieval Crusades, is a lucrative growth industry. "As of 2018, 45 national and 25 subnational jurisdictions are putting a price on carbon," reports the World Bank. "In 2018 the total value of emissions trading systems and carbon taxes is $82 billion," up 50 percent from

2017. Even if climate model projections prove dead wrong and CO_2 is exonerated, the remaining incontrovertible truth is rapidly rising global carbon tax revenues (World Bank and Ecofys 2018).

References

Giaever, Ivar. 2015. "Global Warming Revisited." Lindau Nobel Laureate Meetings. https://www.mediatheque.lindau -nobel.org/videos/34729/ivar-giaever-global-warming -revisited/meeting-2015. Accessed on September 28, 2019.

Jastrow, Robert, William Aaron Nierenberg, and Frederick Seitz. 1990. *Scientific Perspectives of the Greenhouse Problem.* Ottawa, IL: Marshall Press.

Lindzen, Richard S. 1990. "Some Coolness Concerning Global Warming." *Bulletin of the American Meteorological Society* 71(3): 288–299. https://journals.ametsoc.org/doi /pdf/10.1175/1520-0477(1990)071<0288:SCCGW>2.0 .CO;2. Accessed on September 28, 2019.

Lovelock, James. 2014. *A Rough Ride to the Future.* New York: Overlook Press.

Sherwell, Philip. 2011. "War of Words over Global Warming as Nobel Laureate Resigns in Protest." *The Telegraph.* https://www.telegraph.co.uk/news/earth/environment /climatechange/8786565/War-of-words-over-global -warming-as-Nobel-laureate-resigns-in-protest.html. Accessed on September 28, 2019.

World Bank and Ecofys. 2018. "State and Trends of Carbon Pricing 2018." Washington, DC. doi:10.1596/978-1 -4648-1292-7. https://openknowledge.worldbank.org /bitstream/handle/10986/29687/9781464812927.pdf. Accessed on September 28, 2019.

Joel Grossman helped develop research alternatives for the Montreal Protocol climate treaty and writes the Biocontrol Beat blog.

How the Powerful Advance Climate Change
Maxine Gunther-Segal

The dominant ways of understanding the world that are associated with privileged perspectives (white, wealthy, male) are viewed as most legitimate, so that bad ideas and actions—ranging from counterproductive, to violent, to frankly absurd—often dominate simply because they serve the immediate interests of the powerful.

This phenomenon manifests in deadly climate change denialism, which is shockingly pervasive despite the scientific consensus that should make it a fringe perspective. It also manifests in less recognized but more ubiquitous climate change incrementalism: climate policy advocating for gradual reductions in emissions, instead of the immediate transformation needed to avert environmental catastrophe. To push for incremental change in the face of a threat as deadly as climate change is to be in extreme denial. It endangers our entire species and the ecological structures in which we're embedded.

Most of our leaders are complacent climate incrementalists. Understanding how the privileged legitimacy that perpetuates this view works will allow us to recognize and deconstruct it in the real world to say, "I see you using your power and privilege to be believed when you refuse to speak the truth. We won't let you get away with it." How does the legitimacy of knowledge, ideas, and discourse associated with privilege maintain the status quo by facilitating denialism's more "innocuous" yet widespread cousin, incrementalism?

Problems with "Objectivity"

In our culture, distance from a problem is believed to produce more effective solutions. According to this line of thinking, distance minimizes bias so that people are more "objective" and more able to make good decisions. Notice that this idea conveniently happens to serve the most powerful among us. But closeness to a problem makes for better-informed decisions. As feminist theorist Donna Haraway writes, "There is good reason to believe vision is better from below the brilliant space platforms of the powerful" (1988, 583). Decision makers who are close to a problem will be more motivated to be ethical and to incorporate empathy into their response, which are key factors actively rejected by mainstream science.

Being close to a problem can actually make for better governance. There are numerous real-world examples that demonstrate this principle: traditional systems for sharing natural resources like bodies of water, forests, and fisheries (Ostrom 1992). Some systems are thousands of years old, which shows just how sustainable they are. But because they're often created by non-Western and indigenous peoples, they're frequently overlooked by those in the Western world.

Emotionlessness and lack of empathy, prioritized for the sake of "objectivity," support climate denialism. Such feelings promote a lack of urgency in the face of climate change. Emotions allow you to assign importance and meaning to knowledge. In fact, neuroscientist Antonio Damasio has persuasively shown that emotions play a crucial role in decision-making. He states that the bodily changes associated with emotion are "somatic markers" that play a key role, conscious or unconscious, in every choice we make. The knowledge we hold in our bodies, the accumulated wisdom of countless emotional reactions, is crucial to the process of choosing (Damasio 1996).

For instance, Damasio writes about a man he calls Elliot. Elliot's neurological function was totally typical, with the exception of damage to the emotional processing regions of his brain, which made it almost impossible for him to make decisions (Damasio 1994, 95). He writes that "the cold-bloodedness of Elliot's reasoning prevented him from assigning different values to different options, and made his decision-making landscape hopelessly flat" (Damasio 1994, 115). Without access to his emotions, Elliot sometimes spent hours on the smallest decisions.

We can't quibble about minutiae; we have to take immediate action on climate change. But when we're encouraged to deny our emotions and their key decision-making role in the context of an issue like climate change, we're likely to waste precious time on trivial details instead of focusing on structural transformation.

Climate Change: An Opportunity?

By bringing to light the failures of our existing system, climate change presents an opportunity to break down power structures and replace them with something better—something that allows for stakeholders to make decisions and for formerly marginalized perspectives to be respected, instead of the top-down, detached decision-making that brought us the Sixth Extinction. As described by black feminist thinker bell hooks, a marginalized social position "offers the possibility of radical perspectives from which to see and create, to imagine alternatives, new worlds" (hooks 2003, 157). Never before have we so badly needed to imagine new worlds.

These hierarchical breakdowns are already visible in indigenous environmental movements like the fight against the Dakota Access Pipeline at Standing Rock that have pushed back against the hierarchies of legitimate knowledge in the context of colonialism. They have demonstrated to the broader public that indigenous people have an exceptional role and exceptional insight as environmental stewards and defenders.

This breakdown was also visible in February 2019, when young Sunrise Movement activists directly confronted Senator Dianne Feinstein with the consequences of her inaction for their futures. To her face, one protestor said, "Senator, if this doesn't get turned around in ten years you're looking at the faces of the people who are going to be living with these consequences." By acting in defiance of traditional hierarchies, these young activists chipped away at the hierarchy of legitimate perspectives that's leading us toward destruction.

Hope is essential to pushing forward in the face of the horrors of climate change. A kernel of hope exists in the promise that structures like capitalism and patriarchy can be broken down, now that they've been revealed as unable to perform even their original function of maintaining societies that have been superficially functional for elites, while also violently exploitative of the majority. And such structures must break down if we're going to put a stop to climate change before it's too late.

References

Damasio, Antonio. 1994. *Descartes' Error*. New York: Penguin.

Damasio, Antonio. 1996. "The Somatic Marker Hypothesis and the Possible Functions of the Prefrontal Cortex." *Philosophical Transactions of the Royal Society of London, Series B: Biological Sciences* 351(1346): 1413–1420.

Haraway, Donna. 1988. "Situated Knowledges: The Science Question in Feminism and the Privilege of Partial Perspective." *Feminist Studies* 14(3): 575–599.

hooks, bell. 2003. "Choosing the Margin as a Space of Radical Openness." *Framework: The Journal of Cinema and Media* 36: 15–23.

Ostrom, Elinor. 1992. *Governing the Commons: The Evolution of Institutions for Collective Action*. Cambridge, UK: Cambridge University Press. https://wtf.tw/ref/ostrom _1990.pdf. Accessed on September 30, 2019.

Maxine Gunther-Segal is a student at Smith College studying sociology and climate change. She is also a member of the Sunrise Movement, a youth-led organization fighting for climate justice, and aspires to become a lawyer to advocate for climate migrants.

Landslide Mitigation and Perception in Rio de Janeiro
Abigail Hanna

Before my senior year, I spent the summer of 2019 in Rio de Janeiro, Brazil, as an intern working with the NGO Catalytic Communities. CatComm began as a city planning dissertation by Theresa Williamson, founder, and is now a full-fledged network connecting various policies, strategies, and people across the greater Rio area to expand networks and share ideas with an increasing population. CatComm's main focus is on the favelas of Rio, informal settlements mostly situated within the hillsides of the city, and their representation.

My role was to act as a journalist with the newspaper faction of the organization, RioOnWatch.org. I researched landslide and flash flooding events and attended government meetings, learning about potential ways to mitigate and prevent the consequences of these natural disasters. My goal was to publish a timeline of recent landslide and flash flooding events to be utilized and referenced in further CatComm policies and collaborations. In addition to research, I interviewed favela community leaders about their experiences with natural disasters in the area as well as employees of COR, Rio's Operating Center for weather monitoring, and the Defesa Civil, Rio's disaster relief body of government. Speaking with these two different groups of people, I noticed that there were significantly different approaches to climate change debates, depending on who you ask.

A Latin American city facing the ever-intensifying effects of climate change as both a coastal and mountainous region, Rio de Janeiro is an incredibly vulnerable area for water-related disasters. Rather than a longer period of light rain as we may see in the United States east coast, Brazil and its coastal regions in particular receive rainfall in much more intense downpours lasting shorter durations, which can cause water to concentrate very quickly creating land movement or flash flooding. During just one day of Rio's landslide events earlier in April

2019, which killed 34 people total and left hundreds of families homeless, some parts of the city saw over 310 millimeters (12 inches) of rain in 24 hours (Ruge 2019).

When contextualizing climate change in low-income communities in the modern world, one must recognize that the subject should be approached from both a technical and social perspective. The technical side utilizes the tools and technology we have developed to study the weather patterns and natural occurrences that a region is climatically prone to, such as marking repetitions in weather patterns to prepare for another cycle. The social side digs deeper into the anthropocentric effects that have been known to exacerbate these natural occurrences, such as certain political atmospheres that allow wealth and social inequality to flourish.

Whereas a technical response to a given flash flood event would blame a failed siren early warning system or roads that were too full of water to drive down, a social response would blame the population responsible for not mitigating the disaster where it took place: those who were aware of the increasing problems of climate change, yet chose to diminish a budget needed for relief and mitigation to prioritize other financial prospects. For example, during a government meeting last June with Rio's Civil Defense, a member of Rio's CPI or Parliamentary Commission on Flooding and city councilor Teresa Bergher pointed out that no funds were spent on urban drainage systems or hillside containment projects in 2019 (Hanna 2019), a policy aspect of Marcelo Crivella, an evangelical bishop elected by Far Right Brazilian president Jair Bolsonaro for mayor of Rio in 2016. A technical response would blame the government for not providing enough funding; however, a social response blames Crivella specifically, adding a personal layer and providing a face to blame. Technically, it is true that the government is underfunded and therefore less able to provide disaster relief; however, this is due to the fact that the Brazilian government has purposely budgeted less disaster relief funding despite witnessing a growing frequency in water-related disasters in order to serve other, more lucrative purposes, not because relief is particularly expensive or difficult. Social responses perpetuate both negative and positive perspectives: By adding a factor of humanity into the climate change equation, we create methods of analyzing our relationship to nature and the environment. This analysis begs the question, do humans respond to their environments or environments respond to the population changes within them?

Recent academic work has pointed to the former as outdated and the latter as more accurate. If people responded to their environment, this wrongly implies that those most vulnerable to climate change purposely do little to prevent it, when in reality, communities who are mostly at risk for natural disasters are unable to respond properly due to improper infrastructure or lack of funding. From research in climate change conducted over the past century, we know the latter is true: Much of global warming is human-caused and human-exacerbated.

After the 2014 World Cup and the 2016 Olympic Games, Rio shot onto the global stage as a lucrative investment location, with its white sandy beaches and tourist attractions gaining international attention. Like many other Olympic host cities in developing countries with sizable low-income populations, favela residents in Rio are still facing potential eviction even years after the events, even if it is disguised as something else. One of the simplest strategies for landslide and flash flood relief is evacuation. If your house is on a hill and a landslide is coming, you should leave the area. However, due to several trust issues with warning of impending storms and how credible the information being relayed to the population is, many favela residents do not trust those who tell them to evacuate (Melo et al. 2017). Since favelas are informal settlements, the residents do not own the land underneath, so if a house is swept away by a landslide, one would have no choice but to move somewhere else, freeing up that same land for someone (i.e., the Brazilian government) to buy it.

Eviction is veiled as evacuation, and with little to no government help after losing a home, favela residents living in these areas are labeled an environmental problem themselves. So, the question is: How do we decide how much weight social and technical aspects of climate change is held in each community facing issues, and once we do, how can we formulate further climate change policy both by the use of "green" strategies such as proper sewage and recycling systems but further by expanding the definition of "green" to include the interests of low-income communities, to ensure that those communities most affected by climate change will receive the aid they require?

As is obvious in Rio, climate change is happening at an exponential rate, and communities with less opportunity to prevent these natural disasters are the most vulnerable to them. If those with more means do nothing to mitigate these effects, these cultures and communities will not be the only ones at risk: The city of Rio, Brazil, and the world as a whole, could face disastrous environmental consequences.

References

Hanna, Abigail. 2019. "7th Session of Parliamentary Inquiry on Floods Talks Budget Cuts with Rio's Flood Prevention Agency." RioOnWatch.org. https://www.rioonwatch.org/?p=53485. Accessed on September 6, 2019.

Melo, Patricia, et al. 2017. "Evaluation of Community Leaders' Perception Regarding Alerta Rio, the Warning System for Landslides Caused by Heavy Rains in Rio De Janeiro." *National Hazards* 89(3): 1343–1368.

Ruge, Edmund. 2019. "Insufficient Response from Authorities Means Rio Favelas Still Reeling from April Floods." RioOnWatch.org. https://www.rioonwatch.org/?p=53231. Accessed on September 12, 2019.

Abigail Hanna is a senior at Smith College in Northampton, Massachusetts, with a double major in Economics and Portuguese and a concentration in climate change. She enjoys both professional and creative writing as well as playing guitar and hopes to continue working in climate change research and policy.

The Little Big Changes—How Big Is Your Backyard?
Hogyeum Joo

I received my Bachelors in Ecosystems and Human Impact (Sustainability Studies Program) in May 2019 from Stony Brook University. It's a multidisciplinary program that covers a variety of subjects in sustainability—from economics and environmental chemistry to anthropology and ecology. As I planned my courses, my advisors often asked me and my other peers "What do you want to do with this major? Why are you in this program?" Some answers from my peers were simple: "to get a job at a company" or "I want to work for this nonprofit." But I always had a hard time answering that question. I had so many dreams, I just didn't know how to answer it. My usual answer was "well . . . I'm not sure. But I want to do something to change the world. I just don't know how to do that yet." My answer was mostly about "how" than "what."

Climate change is a concept that's familiar to most of the people these days, no matter what age you are or what you do for your living. When it comes to climate change, people tend to think about the "big" issues the media covers—melting glaciers and suffering polar bears, the Amazon forest on fire, hurricanes, flooded coastal zones, etc. However, in this essay, I'd like to talk about the "small" changes we can make to solve these big issues.

I do understand and value the need and significance of macroanalysis on certain issues such as global climate change trend or ocean temperature changes. Actions on those issues usually take place at much larger scales by more powerful people with greater powers. However, I think the changes in current environmental conditions can happen in better ways if we focus more on smaller issues.

I always believed that there are two ways to change the world—through societal-level changes and through education. Societal-level changes require changes in laws and regulation. Enforcement of regional/national/global policies can make a huge difference. However, it also requires many preconditions—educated voters, responsible elected officials who are willing to work for their voters, policies that can work despite the disagreement from certain interest groups, and most importantly, people's will to make changes. Our societies have achieved many things through this way so far. We achieved racial equality, women's right to vote, etc. Leaders of the world carried out agreements such as the Tokyo Protocol and the Paris Agreement on climate change this way.

However, changes and people's passion toward change never stopped growing. The main driving force behind this was education. The *Merriam-Webster Dictionary* defines the word *educate* as "to develop mentally, morally, or aesthetically especially by instruction." People have been teaching what is right and just at schools and at home. People in different generations developed their ideas, grew their passion, and acted based on their beliefs. This process allowed the citizens to show their opinions through our election system, legal processes, and policy-making procedures. The cycle of changes and education continues. I believe it's very simple and basic like this when it comes to environmental education as well.

Education is more than the kind that happens in classrooms. Education happens when parents take their children out to go camping, or even when a child learns how to put trash in a trashcan or when they start to learn how to appreciate what they have around them. I believe those "big" decisions and actions are ultimately for those who are around us. Look around you. Nothing you'll see is new. Your family, trees on the streets, and the birds singing by the window. It will be all gone if we don't do anything. However, how often do we pay attention to these when we talk about climate change?

Different people take climate change issues differently. Some care a lot more than others, and some don't necessarily care about it at all. Some people see themselves as global citizens and fight for the deforestation issue of the Amazon in New York. Some people see themselves as just simply one of the 7.7 billion people on this planet who can't make any differences in this world. However, one thing we're all forgetting here is that we all share the responsibilities and impacts of the results together.

I'd like to ask one question to the readers. How big is your backyard? Your backyard can be the Amazon forest, or the entire Pacific Ocean, or the size of your actual backyard. This is a question about ourselves and our future. To what extent, can we make changes by taking actions in this world?

I'm a student who studies sustainability from a variety of perspectives. Scholars and students often make a mistake by thinking that everyone should know everything about the environment because it's important. My opinion is a little different. I believe that not everyone has to be an ecologist or environmental scientist to do his or her part on this issue.

I believe that the big changes can always start from smaller thoughts—as small as the little tree that's growing in your backyard. I strongly believe that if we could create the culture that can make children at home appreciate the tree shades in their own backyards and make them wonder what they can do for the trees to pay back to them, that'd be a great step forward for the humanities and for the environment. When we can make everyone look at the world around them this way, we will be able to change the world and our societies and people, region by region and one by one.

Personally, I'm still not entirely sure if I can answer to myself about how I want to change the world. But I have confidence in one thing for sure. Doing what I can do in the best way possible is the best thing I can do. After reading this, look around yourself, and ask yourself the same question. How big is my backyard? How are you going to change the world?

Hogyeum Joo recently graduated from SUNY Stony Brook's Sustainability Studies Program with honors (BA in Ecosystems and Human Impact). With his passion toward environmental sustainability and environmental awareness, Hogyeum is currently pursuing his master's degree in Sustainability Management at Columbia University in the City of New York. He also serves as an Alumni Representative for the Friends of Ashley Schiff Preserve, a community-based environmental organization based in Stony Brook, New York.

Give Me Money, Give Me Power, and Give Them Death
Joy Semien

Climate change does not impact all communities equally. It affects people of color and those in lower income brackets more seriously than it does whites and the middle and upper classes. In this respect, now, as in the past, people of color and poor people suffer disproportionately from environmental harms resulting from industrial development.

Inhale and now exhale. What did you smell? Did you smell fresh air or did you smell a strong potent odor that knocked you off your feet? Every day, hundreds of thousands of community members living near heavily industrialized facilities awaken to the smell of a heavy stench of cancer-causing toxins also thought of as the "smell of money."

I remember as a child, every morning waking up with my sister to the smell of grandma's breakfast and the good ole "smell of money." As a curious child, I would regularly ask if our impoverished family would ever see the money we smelled. The response was always the same, "oh child" that money isn't for us" and I would respond, "well then I don't want to smell it." My grandmother would laugh and continue cooking breakfast, but that curiosity led me to explore the truth behind the scent I smelled daily.

As I matriculated in age, I soon learned that there were thousands of communities across the United States awakening to the same smell of money. These communities suffered many of the same health effects that my community regularly faced and were referred to as Environmental Justice (EJ) communities (Bullard 2000). These communities often consisted of poor, people of color with little resources to fight big industries. Like my community, residents were practically voiceless; they lacked power, resources, and the training needed to prevent big industry from destroying their livelihoods with harmful pollutants. The lack of these traits set them up to be mistreated by those who were far wealthier and much more powerful. In many cases, the wealth and power were linked to the advancement of political agendas at the cost of the residents living in these communities. Since the residents lacked a voice, they often went unnoticed and eventually died (Allen 2003).

As a little girl growing up in one of these communities, I watched firsthand as family members were ravaged by cancer, Alzheimer's disease, asthma, and a host of other health issues known to be linked to the toxins released by these industrial facilities (Reed 1991). The small, unincorporated community that I grew up in is Geismar, Louisiana, located in the heart of the Louisiana Chemical Corridor also known as "Cancer Alley" (Wright 1998). Geismar, initially developed as a small postal outpost along the southern railroads, is conveniently located across from the Mississippi River. As a result of its location, having access to water and rail transportation, the community colonizers—two German brothers—built a sugarcane plantation that housed hundreds of slaves (Sternberg 2013). Post-Emancipation Proclamation, these same slaves became sharecroppers and eventually earned enough money to purchase the land on which generations before them worked (Allen 2003). However, little did they know in a few short years they would be bombarded with a new type of injustice that would again strip them of their newfound power and wealth. As the new "slave masters" came into the community, they brought with them tens of thousands of pounds of pollutants that would be released in the air and water slowly killing generations of descendants of African slaves who were wrongfully brought to the America's to build a community that only wanted them for their blood and sweat (Allen 2003).

It wasn't until the environmental justice movement really took off in the mid-1960s, that people began to pay attention to communities like Geismar. Community leaders began to rise up out of the pits they were living in, join hands and unite creating movements like the sanitation strike of 1968 and the Warren County sit-in of 1982 (Bullard 2014). These movements helped communities like Geismar gain a voice and forced governments to create, change, and reevaluate policies (Bullard and Johnson 2000). For many communities who gained a voice this opportunity was great, but for other communities that endured historical marginalization as a result of racism and pure hatred, these communities have remained voiceless while CEOs, politicians, and even research institutions continued to capitalize off their environmental injustices. In other words, the wealthy grew wealthier, the poor grew poorer, and the empowered gained even more power. Even today, communities like those in the Louisiana Chemical Corridor are still faced with these injustices as new petrochemical plants are being built in small towns with a radius of less than 10 square miles like Geismar (Wright 1998).

Although the direct impacts for those living in these communities are long-term health impacts, these same injustices indirectly impact the rest of the world. The Intergovernmental Panel on Climate Change (IPCC) has determined that 95 percent of all human activities, inclusive of industrial pollutants, has been linked to the warming of the Earth's temperature by over 0.85°C (IPCC 2013). Scientist have predicted that the changes in the climate have a direct relationship with the increase in natural disasters across the world like heatwaves, tsunamis, floods, tornadoes, fires, and even hurricanes. According to the IPCC, lowering the Earth's temperature by 1.5°C to 2°C can play a role in reverting some of the detrimental consequences already present on the Earth today. This reduction can mostly be attributed to reducing the amount of pollutants humans contribute to the Earth's atmosphere on a yearly basis. For communities that are already vulnerable, decreasing the Earth's temperature is key to protecting their health, safety, and welfare.

In recent years, politicians have encouraged the implementation of buyback programs as an effort to reduce regulations. However, in this scenario industries are ultimately left to an honor system for emission reduction, which in many cases goes unchecked or operates with impunity. As a result, communities are still left voiceless, as they intake deadly pollutants (Wright 2012). It is time to give a voice to the voiceless and return power to the powerless, while building communities that are more sustainable, as well as, resilient against both natural and anthropogenic disasters.

References

Allen, Barbara L. 2003. *Uneasy Alchemy: Citizens and Experts in Louisiana's Chemical Corridor Disputes*. Cambridge, MA: MIT Press.

Bullard, R. D. 2000. *Dumping in Dixie: Race, Class, and Environmental Quality*. 3rd ed. London: Routledge.

Bullard, Robert D., and Glenn S. Johnson. 2000. "Environmental Justice: Grassroots Activism and Its Impact on Public Policy Decision Making." *Journal of Social Issues* 56(3): 555–578.

IPCC. 2013. "IPCC Assessment Report Summary for Policy Makers." https://www.ipcc.ch/site/assets/uploads/2018/02/WG1AR5_SPM_FINAL.pdf. Accessed on September 24, 2019.

Reed, Susan. 1991. "His Family Ravaged by Cancer, an Angry Louisiana Man Wages War on the Very Air That He Breathes." *People*. https://people.com/archive/his-family-ravaged-by-cancer-an-angry-louisiana-man-wages-war-on

-the-very-air-that-he-breathes-vol-35-no-11/. Accessed on September 24, 2019.

Sternberg, Mary Ann. 2013. *Along the River Road: Past and Present on Louisiana's Historic Byway.* Baton Rouge: Louisiana State University Press.

Wright, Beverly. 1998. "Endangered Communities: The Struggle for Environmental Justice in the Louisiana Chemical Corridor." *Journal of Public Management and Social Policy* 4(2): 181–191.

Joy Semien holds a Master's degree in Urban Planning and Environmental Policy from Texas Southern University and a Bachelor's degree in Biology from Dillard University. Joy has studied abroad and conducted research in places like Paris, France; Gabon, Africa; and Kingston, Jamaica. Joy has received awards for both her research as well as her extensive community service along the Gulf Coast from organizations like the American Planning Association, Planning in the Black Community Division, and the Louis Stokes Louisiana Alliance for Minority Participation. Joy's research interests are in developing methods to increase the capacity of residents within multi-hazard communities of color and of low income to prepare, respond, and recover from disasters.

No book on climate change is complete without some mention of the individuals and organizations who have been involved in the history of this topic. As with most subjects in science, the slow, irregular evolution of ideas and information about global warming and climate change is essential to our present-day understanding of the topic. Such a history illustrates the ways in which starts and stops, misleading ideas, and serious controversies over the centuries have finally led to our current understandings of the factors that affect climate change and, in particular, the way human activities have also affected Earth's changing climate. This chapter provides brief summaries of a few of the most important individuals and organizations that had played a role in this story.

Svante Arrhenius (1859–1927)

Among his many scientific contributions, Arrhenius is generally credited with being the first person to suggest that carbon dioxide released by human activities was sufficient to change the concentration of carbon dioxide in the atmosphere and, thereby, to affect the planet's annual average temperature. His research on global warming stands as one of the primary accomplishments of today's modern science of climate change. In

16-year-old climate activist Greta Thunberg being interviewed during a demonstration outside the Swedish Parliament House, Riksdagshuset, on February 15, 2019. (Per Grunditz/Dreamstime.com)

addition to his work on global warming, Arrhenius is known today as one of the founders of the science of physical chemistry, a field in which the principles of physics are applied to studies of chemical phenomena.

Svante August Arrhenius was born in the town of Wijk (also Wik or Vik), near Uppsala, Sweden, on February 19, 1859. He was considered to be a child prodigy who taught himself to read at the age of three. At the age of eight, Arrhenius entered the local grammar school as a fifth grader. There he became especially well known for his skills in mathematics and physics. He graduated with highest honors in his class in 1876. He then entered the University of Uppsala, with plans to major in physics. Dissatisfied with the academic offerings there, he soon transferred to the Physical Institute of the Swedish Academy of Sciences in Stockholm. There he focused his research on the conductivities of electrolytes, a subject on which he eventually wrote his doctoral thesis.

This topic was of some interest because it melded Arrhenius's interest in both physics and chemistry. The thesis was later to form the basis for his receiving the 1903 Nobel Prize in Chemistry. The thesis was not well understood or supported by the faculty at Uppsala, at least partly because the application of physics to chemistry was a somewhat new and poorly understood field of science. In fact, Arrhenius's work in his across disciplines was one reason that he later found it difficult to advance through his academic career, often encountering opposition from his colleagues.

In 1891, Arrhenius was offered a position as lecturer at Stockholm University College (Stockholms Högskola, now Stockholm University). He was promoted to professor in 1895 and was named rector of the university a year later. He remained in this position until his retirement in 1905. He was then offered a position as director of the newly created Nobel Institute for Physical Chemistry in Berlin, a position he held until his death in Stockholm on October 2, 1927.

Arrhenius was interested in a variety of topics beyond physical chemistry. He conducted original research on topics as diverse as the application of physical chemistry to the study of toxins and antitoxins, origin of the ice ages, birth of the solar system, and properties of comets and other extraterrestrial phenomena. Among the many honors and awards he received during his lifetime in addition to the Nobel Prize were the Davy Medal of the Royal Society of London, the Willard Gibbs Award of the Chicago section of the American Chemistry Society, the Faraday Lectureship Prize of the Royal Chemical Society, the Franklin Medal of the American Philosophical Society, and election as a foreign member of the Royal Society. He also received honorary doctorates from the universities of Birmingham, Cambridge, Edinburgh, Greifswald, Groningen, Heidelberg, Leipzig, and Oxford. Among the many popular books he wrote were *Världarnas Utveckling* (Worlds in the making), *Stjärnornas Öden* (Destiny of the stars), *Smittkopporna Och Deras Bekämpande* (Smallpox and its combating), and *Kemien Och Det Moderna Livet* (Chemistry and modern life).

Center for Climate and Energy Solutions

The Center for Climate and Energy Solutions (C2ES) was founded in 1998 as the Pew Center on Global Climate Change. The Center was one of a group of research centers created in 1990 by the Times Mirror Company as a service for conducting public opinion polls on politics and policy. The Center evolved over time to have separate units focused on U.S. Politics and Policy, Journalism and Media, Internet and Technology, Science and Society, Religion and Public Life, Hispanic Trends, Global Attitudes and Trends, Social and Demographic Trends, and Research Methodology. The purpose of the organization is "to advance strong policy and action to reduce greenhouse gas emissions, promote clean energy, and strengthen resilience to climate impacts." It operates on the assumption that the development of a strong, cost-effective economic policy is the best way to deal with problems of climate change. In 2011, the organization adopted a new name, Center for Climate and Energy Solutions, with essentially the same goals and objectives as the original group.

At the time of its founding, the Center also created the Business Environmental Leadership Council, consisting of 13 businesses involved with or concerned about the role of energy use in global warming. Today, that list of companies has grown to 35 and includes such well-known businesses as Alcoa, Amazon .com, Bank of America, Dow, General Electric, Goldman Sachs, Honda, IBM, Microsoft, Shell, and Toyota. The companies have financial assets of an estimated $3 trillion with more than 3.7 million employees. In 2011, the companies accepted a set of guiding principles that included an acceptance of scientific consensus about climate change and its potential effects on the environment and the human race, businesses should accept responsibility for finding ways to reduce greenhouse gas emissions, the United States must adopt a set of actions to meet international greenhouse gas emission goals, and climate change is a global problem that requires the attention of developed and major developing economies.

C2ES points to a number of important accomplishments during its history. These include:

- Working with Senators John McCain and Joseph Lieberman to develop a bipartisan bill to develop a market-based climate program
- Sponsorship of the Climate Dialogue at Pocantino in Tarrytown, New York, and involving 15 nations meeting to discuss ways of dealing with climate change
- Issuance of a report, "Adapting to Climate Change," outlining the role of the federal government in adopting policies and practices designed to strengthen the economy and natural resources in the face of climate change
- Co-founding the Carbon Capture Coalition to develop policies and practices for the capture of carbon dioxide as a mechanism for reducing greenhouse gas emissions to the atmosphere

- Founding of the Solutions Forum, a nationwide group of leaders from business and state and local government for finding ways of reducing greenhouse gas emissions and adapting to change likely to develop as a result of climate change
- Launch of Climate Innovation 2050, an initiative designed to find ways of "decarbonizing" the American economy and moving toward a sustainable society with reduced use of fossil fuels

The products of the many C2ES programs are available on the organization's website. They include webinars, reports of research, blog posts, videos of live events, policy maps, position papers, fact sheet, and other resources on market-based approaches to dealing with climate change. Some of the topics for which resources are available are International Civil Aviation issues, a quick review of possible climate plans, congressional carbon-pricing proposals, green solutions to the Midwest's growing flooding problem, state carbon pricing policies, city climate policies, local action on energy efficiency, greening city transportation fleets, examples of energy-efficient cities, public-private collaboration on city climate resilience planning, and mayors' roles in dealing with climate change.

The C2ES also has a very useful collection of basic information on the causes, patterns, and effects of climate change. The major divisions of this web page deal with climate science, climate impacts, extreme weather, energy/emissions data, and the climate classroom. The last category, as an example, has detailed information on climate basics for children, teachers' resources, a section on "What We Can Do," and classroom lessons on climate change.

Climate Action Network

Climate Action Network (CAN) was founded in 1989 at a meeting in Hanover, Germany, of about 20 activists from Germany and the United States. The group had become convinced about the reality of climate change and met to discuss ways in which those with similar views could join and work together to deal with the issue. The organization has grown over the years and now consists of more than 1,300 nongovernmental organizations in over 120 countries. These groups are organized into two regional networks, East Africa and South Africa, and 10 national groups, located in Australia, Canada, China, France, Japan, New Zealand, South Africa, Tanzania, Uganda, and the United States. Some of the member organizations are (R)Evolution Let's Change Now (India), Acción Ecológica (Chile), ActionAid (United States), Arab Youth Climate Movement (regional), Association of Voluntary Actions for Society (Bangladesh), Campaign against Climate Change UK (United Kingdom), Canopée (France), CARE International Vanuatu (Vanuatu), Diaries of the Ocean (Lebanon), Environmental Law Center Armon (Uzbekistan), Global Environment Centre (Malaysia), Huvadhoo Aid (Maldives), Náttúruverndarsamtök Íslands / Iceland Nature Conservation Association (Iceland), Khazer—Ecological and Cultural NGO (Armenia), and Oil Change International (United States).

The primary goals of CAN are exchange of information among member groups and coordinated development of programs for dealing with climate change. The latter goal is achieved through a series of working groups in which individual interested organizations join together to discuss and develop action plans on specific aspects of the climate change crisis. As of early 2020, working groups exist for the following topics: Adaptation and Loss and Damage, Agenda 2030, Agriculture, Ambition and NDCs, Bunkers, Comms, Ecosystems, Energy, Finance, Flexible Mechanisms, G20, Global Stocktake, Long-Term Strategies and Climate Action Initiatives, Mitigation, NGO Participation, Science Policy, Short Lived Climate Pollutants, Technology, and Transparency. As an example, the Bunkers Working Group focuses on pollution caused by international aviation and oceanic and land shipping. Since these activities typically involve the movement of goods and people across country borders, carbon costs cannot be assigned to any individual nation, and this source of greenhouse gas emissions is often omitted from regional and international agreements on climate change. This working group considers existing treaties and other agreements, as well as those currently being discussed, with recommendations from the working group to the enabling agencies.

In addition to working groups, CAN also sponsors larger, more widespread campaigns on topics of broad interest. The two campaigns currently underway are The Big Shift Global and Implementing Low Carbon Development. The former campaign is aimed at encouraging the World Bank to adjust its policies to reflect the goal of the IPCC and other organizations to keep global temperature increases to 1.5°C by 2020. The latter campaign is designed to find ways to reduce climate change while promoting programs of sustainable development at the same time. More information on these campaigns can be found on the CAN website at http://climatenetwork.org/campaigns.

In addition to an excellent "Media" section on the CAN website, providing a variety of resources for members and the general public, the organization also publishes an online newsletter, ECO, and maintains a blog with the same name. An ECO app can also be downloaded from the Apple Store or Google Play.

Eunice Newton Foote (1819–1888)

Foote is one of those less well-known researchers in the history of science who made important discoveries that were largely ignored or forgotten. Arguably, to no small extent, this lack of recognition resulted from Foote's having been a woman attempting to announce a scientific breakthrough in a culture in which feminine intelligence and skills were largely discredited.

Foote's premier accomplishment was the completion of a paper on "Circumstances Affecting the Heat of the Sun's Rays." (The original paper can be found online at https://static1

.squarespace.com/static/5a2614102278e77e59a04f26/t
/5aa1c3cf419202b500c3b388/1520550865302/foote
_circumstances-affecting-heat-suns-rays_1856.pdf.) For whatever reason, she did not announce the paper in person herself
but had it read to the American Association for the Advancement of Science (AAAS) at its annual convention in 1856. In
her short (two-page) paper, Foote described a series of experiments in which she attempted to determine the effect of the
Sun's rays on the temperature achieved by various gases. The
design of the experiment was simple. She exposed a series of
test tubes containing various gases and/or concentrations of
gases to sunlight and measured the temperature of the trapped
gas in each case. She concluded that the greatest temperature
rise occurred in the tube containing carbon dioxide (or, as it
was known as the time, carbonic acid, the water solution of
carbon dioxide). The conclusion she drew from her research
was that

> An atmosphere of that gas [carbon dioxide] would give
> to our earth a high temperature; and if as some sup-
> pose, at one period of its history the air had mixed with
> it a larger proportion than at present, an increased tem-
> perature from its own action as well as from increased
> weight must have necessarily resulted. (https://static1
> .squarespace.com/static/5a2614102278e77e59a04f26/t
> /5aa1c3cf419202b500c3b388/1520550865302/foote
> _circumstances-affecting-heat-suns-rays_1856.pdf.
> Accessed on December 2, 2019)

This discovery was essentially the same later announced by
Irish physicist John Tyndall in about 1859. Tyndall's experiments were somewhat more advanced than Foote's, although
his conclusions were similar to hers.

Eunice Newton was born on July 17, 1819, the eldest of
12 children of Thirza and Isaac Newton Jr., of Goshen, Connecticut. She attended the Troy Seminary for Women, in Troy,

New York, from 1836 to 1838. During that period, she was also allowed to attended classes in science taught nearby by the progressive educator Amos Eaton. Eaton was one of the most influential proponents of the time for the inclusion of science in a person's general education. He also held the (at the time) revolutionary idea that such an education ought to be as readily available to women as it was for men. The atmosphere in which Newton found herself at the time was, to say the least, unusual for women of the day.

Newton also benefited from a somewhat forward-looking policy at the AAAS, which allowed women to become members. The association continued to hold at least partly to a sexist policy, however, as it created categories of "professional" and "fellow" for male members, while women were described simply as "members." Women were also prohibited from presenting papers at AAAS meetings, a restriction that forced Newton to have her paper read to the congress by American chemist Benjamin Silliman. The paper was also excluded from the printed annual records of reports presented to the association during the previous year.

Newton had married Elisha Foote, a judge and amateur researcher himself, in 1841. Elisha Foote was interested in the properties of gases, a topic on which he presented a paper before the AAAS. In addition to her interests in science, Eunice was also a committed suffragette who was active in the famous Seneca Falls Convention for women's rights in 1848. She was one of five women who served on the publication committee for the convention and, along with her husband, was a signatory to its Declaration of Sentiments. She is not known to have produced any other scientific works during her lifetime. She is described in her genealogical record, however, as "a fine portrait and landscape painter . . . an inventive genius, and a person of unusual beauty" (Newton Genealogy. 2019. http://www.ebooksread .com/authors-eng/ermina-newton-leonard/newton-genealogy -genealogical-biographical-historical-being-a-record-of-the --noe/page-110-newton-genealogy-genealogical-biographical

-historical-being-a-record-of-the--noe.shtml. Accessed on September 29, 2019).

Foote died at her home in Lenox, Massachusetts, on September 30, 1888.

Al Gore Jr. (1948–)

Every social movement has a handful of charismatic and outspoken leaders who carry the message of change to the general public. For the civil rights movement of the 1960s, for example, one of those individuals was Reverend Martin Luther King Jr. No one person can "carry" a movement, of course, and several other names can be associated with the civil rights and other important social movements. In the field of climate change, however, at least two names stand out among those individuals who have framed the problem for the general public, explained the risks and challenges involved, and suggested ways in which progress can be made in the field: Al Gore Jr. and James Hansen (see next profile).

Gore's interest in climate change dates to his years as a student at Harvard University in the early 1960s. Among the courses he took there was one on global warming, taught by oceanographer Roger Revelle, a pioneer in studies of the fate of carbon dioxide in Earth's carbon cycle. After Gore was elected to the U.S. House of Representatives in 1976, he continued to think about the fate of Earth's climate and the steps humans needed to take to deal with this developing problem. Five years later, he convened what is said to be the first congressional hearing on climate change. He later called similar hearings in 1982 and 1984. In 1985, a year after he was elected to the Senate from Tennessee, Gore also called for a designated "Year of the Greenhouse" modeled after the 1957–1958 Geophysical Year. (The recommendation was never followed up on by politicians or scientists.)

Eight years later, presidential candidate Bill Clinton chose Gore to run as his vice-presidential candidate in the election of

1992, a campaign in which they were successful. In the same year, Gore published a book on climate change for which he is probably best known, *Earth in the Balance: Ecology and the Human Spirit.* The book was updated and reissued in 2006, the same year in which Gore's film *An Inconvenient Truth* was released. That film won the Academy Award for the Best Documentary a year later. In 2007, Gore was also awarded a share of the Nobel Peace Prize (along with the Intergovernmental Panel on Climate Change) for his efforts on behalf of public education on global warming and climate change.

Albert Arnold Gore Jr. was born in Washington, D.C., on March 31, 1948. His parents were Pauline (LaFon) Gore, an attorney, and Albert Arnold Gore Sr., a member of the U.S. House of Representatives from 1939 to 1953 and the U.S. Senate from 1953 to 1971. Gore attended St. Albans School, an independent college preparatory day and boarding school for boys in Washington, D.C., from 1956 to 1965. During the summer, he worked on the family tobacco and hay farm in Carthage, Tennessee. Upon his graduation from St. Albans, Gore entered Harvard University, with plans to major in English. He later changed his major to government, a field in which he earned his Bachelor of Art degree in 1969.

At that point, Gore faced the difficult choice, in the midst of the Vietnam War, of joining the military or, as many of his classmates decided, to find a way of avoiding being drafted. He chose the former option and enlisted in the army in August 1969. He was eventually deployed to Vietnam in 1971, where he served until he received his honorable discharge in May of that year. Upon his return to the United States, he enrolled at the Vanderbilt University Divinity School in a program designed for secular careers. During his evenings, he also worked as an investigative reporter for the *Tennessean* newspaper.

In 1974, Gore changed the direction of his career interests and enrolled at the Vanderbilt University Law School. He continued his studies there until 1976 when he decided to run for the newly vacant seat in the 4th district of Tennessee. Gore won that seat and was reelected three more times, in 1978, 1980, and 1982. In 1984, he then ran for the U.S. Senate from Tennessee, an election he won. He was then reelected to the Senate in 1990, before joining Bill Clinton's ticket in the 1992 presidential election. Gore served as Clinton's vice president from 1992 to 2000. In the two decades since he left the political arena, Gore has devoted his time and efforts primarily to educating the general public about critical environmental issues, most importantly, global warming and climate change.

James Hansen (1941–)

James Hansen is probably best known today for an appearance he made on June 23, 1988, before the U.S. Senate Committee on Energy and Natural Resources. On that occasion, Hansen began his testimony by saying that

> Number one, the earth is warmer in 1988 than at any time in the history of instrumental measurements. Number two, the global warming is now large enough that we can ascribe with a high degree of confidence a cause and effect relationship to the greenhouse effect. And number three, our computer climate simulations indicate that the greenhouse effect is already large enough to begin to effect the probability of extreme events such as summer heat waves. (Hansen's complete testimony can be found at https://www.sealevel.info/1988_Hansen_Senate _Testimony.html#targetText=On%20June%2023%2C %201988%2C%20James,on%20Energy%20and%20 Natural%20Resources)

This testimony was certainly not the first time that scientists had spoken out about the risks of possible global warming. But, for whatever reason, Hansen's observation caught the attention not only of the committee members before he appeared but also of reporters from print and electronic resources and the general public as a whole. The testimony did not set off a flurry of activity for dealing with climate change. But it is probably the first time at which people at all walks of life became aware of the threats posed by global warming and the needs to begin doing something about those threats.

The Senate committee testimony was by no means the last time that Hansen was to make his case about global warming before other scientists, policy makers, and the general public. In fact, he has spent much of the last 30 years carrying this message in speeches, articles, seminars, and any other setting at which people would listen to him. If there is anyone today who is and has been carrying the gospel of climate change awareness, it is probably James Hansen.

James Edward Hansen was born in Denison, Iowa, on March 29, 1941. He enrolled in the space program at the University of Iowa from which he earned his BA in physics and mathematics (with highest distinction) in 1963, his MS in astronomy in 1965, and his PhD in physics in 1967. He spent the academic year of 1966–1967 as a visiting student at the Institute of Astrophysics at the University of Kyoto and the Department of Astronomy at Tokyo University. He then began a long career associated with the Goddard Institute for Space Studies at Columbia University in New York City. That relationship was to last until 2013, during which time he held a variety of titles, including director of the institute from 1981 to 2013. In 1969, Hansen was also appointed a research associate at Columbia, where he has remained ever since. He is currently director of the Program on Climate Science, Awareness and Solutions at Columbia.

Hansen is author or co-author of well over 150 professional publications on the physics and chemistry of the atmospheres of Venus and Earth, as well as a host of topics in the field of climate science. He has received more than four dozen honors and awards including the National Wildlife Federation Conservation Achievement Award (1988), Goddard Space Flight Center William Nordberg Achievement Medal (1996), John Heinz Environment Award (2001), Roger Revelle Medal of the American Geophysical Union (2001), Leo Szilard Award of the American Physical Society (2007), Rachel Carson Award for Integrity in Science of the Center for Science in the Public Interest (2008), Carl-Gustaf Rossby Research Medal of the American Meteorological Society (2009), Blue Planet Prize of the Asahi Glass Foundation (2010), Steve Schneider Climate Science Communications Award (2012), Walker Prize of the Museum of Science of Boston (2014), and Tang Prize in Sustainable Development (2018).

Syukuro Manabe (1931–)

Manabe's special area of expertise is the modeling of factors that may affect climate change. His earliest research in the field dates to the late 1950s, before powerful computers were available. As computer hardware has improved over the past six decades, so have the forecasts produced by general circulation models, of which Manabe is a master. His recent research has focused on the effects of clouds on climate change, projected changes in soil moisture as a result of global warming, and the interaction of atmosphere and oceans in response to increased levels of carbon dioxide.

Syukuro Manabe was born in Shingu, Shikoku-Chuo-Shi, Ehime-Ken, Japan, on September 21, 1931. He received his Bachelor of Science (1953), Master of Science (1955), and Doctor of Science (1959) degrees from Tokyo University. He then moved to the United States, where he took a position as research meteorologist in the General Circulation Research Section of the U.S. Weather Bureau. He held that post until 1963, when he transferred to the Geophysical Fluid Dynamics Laboratory of the National Atmospheric and Oceanic Administration, where he held the post of senior research meteorologist. In 1997, he accepted an appointment as director of the Global Warming Research Program at the Frontier Research Center for Global Change in Tokyo, Japan. In 2002, he moved to the Japan Marine-Earth Science and Technology

Organization in Kanagawa, Japan, as a consultant, a post he held until 2009.

Throughout this long period, Manabe has also held a series of academic appointments, beginning with his appointment as lecturer, with rank of professor, in the Program in Atmospheric & Oceanic Sciences (PAOS) at Princeton University. He remained in that post until 1997. He returned to Princeton in 2002, when he was named visiting research collaborator at PAOS, and again in 2005, when he was appointed senior meteorologist at PAOS, a title he continues to hold today. During the first two decades of this century, Manabe also served as visiting professor at the School of Environmental Sciences and the University of Nagoya, Japan.

Manabe has received a number of honors, including the Gold Medal Award of the U.S. Department of Commerce in 1970, Blue Planet Prize of the Asahi Glass Foundation in 1992, Asahi Prize of the Asahi News-Cultural Foundation in 1995, Volvo Environment Prize from the Volvo Environment Prize Foundation and Minister's Award from the Japan Ministry of Environment, both in 1997, Benjamin Franklin Medal of the Franklin Institute in Philadelphia in 2015, Frontiers of Knowledge Award of the BBVA Foundation in 2016, and the Crafoord Prize of the Royal Swedish Academy of Sciences in 2018. Manabe also received an honorary doctor of science degree from McGill University in 2004 and was inducted into the Earth Hall of Fame in Kyoto, Japan, in 2009.

Past Global Changes

Past Global Changes (PAGES) is part of a large, ongoing, international program designed to produce a better understanding of the relationship among the biological, chemical, physical, and human components of planet Earth. One of the earliest manifestations of this effort was the International Geosphere-Biosphere Programme (IGBP), founded by a group of interested researchers in 1987. That organization was later involved with the creation of other international groups with similar or related concerns, including the International Panel on Climate Change in 1988, the Earth Summit in 1992, and the International Human Dimensions Programme on Global Environmental Change in 1996. Among the most recent of these organizations is Future Earth, created in 2012 as a way of strengthening the interface between science and policy. (An excellent overview of this somewhat complex story can be found at "Towards Future Earth: Evolution or Revolution?" 2015. Global IGBP Change. http://www.igbp.net/news /features/features/towardsfutureearthevolutionorrevolution.5 .950c2fa1495db7081e18780.html. Accessed on September 29, 2019.)

PAGES has a long connection with these organizations, dating to 1991. Within the general field of planetary environmental interactions and change, it has focused its attention on past history, dating from the Pliocene epoch to the present day. The organization is interested in discovering all aspects of the interaction among Earth's environment, its climate, and the role of human activities within this system. The organization was originally funded by the U.S. and Swiss National Science Foundations. It is currently supported by the Swiss Academy of Sciences and the University of Bern, Switzerland.

PAGES' work is organized around nine major areas of focus:

- Facilitating international research on past environmental changes
- Promoting synthesis of scientific knowledge and data
- Strengthening the involvement of scientists from developing countries
- Integrating the paleoscience and wider global change communities
- Disseminating research findings and organizational information
- Supporting scientific training and education

- Integrating scientific evidence from observations and modeling
- Ensuring public access to paleoscientific data
- Enhancing the visibility and use of paleoresearch

Membership in PAGES currently consists of more than 5,000 researchers in over 125 countries. Membership information and forms are available on the PAGES website at http://pastglobalchanges.org/people/people-database/join-pages. The organization achieves its objectives through temporary working groups that focus on specific topics of interest. Examples of such working groups are those dealing with sea-level changes, forest dynamics, peat, ocean carbon and circulation, fire, land cover, volcanoes, and interglacial periods. Results of these studies are made generally available to professionals in the field as well as the general public through a variety of means. The Highlights section of the PAGES website lists some of the most important studies that have been completed, along with a searchable database of the complete bibliography. Other resources are journal articles on PAGES research, as well as special issues of regular journals.

Much of the most important research findings produced by PAGES teams can be found in the *PAGES Magazine*. All articles in all issues of the magazine are available free of charge for downloading or in hard-copy format. Recent topics of the magazines have been sea-level rise, past land use and land cover, Earth's biodiversity, and climate change and cultural evolution. In addition to its print and electronic resources, PAGES makes available reports from its workshops and other meetings. They are available in the form of abstracts, full reports, posters and presentations, and other types of meeting products on the organization's website at http://pastglobalchanges.org/products/meeting-products. In addition, PAGES provides several other types of educational material for professionals and the general public. These resources include a monthly newsletter, "e-news," articles and flyers, plans and strategies developed by working groups, posters and presentations, PowerPoint slide presentations, and press releases.

PAGES offers a variety of meeting types to meet goals of participants. The largest and most inclusive is the quadrennial Open Science Meeting, at which participants can present their own research and hear about developments in their own and related fields. They generally attract anywhere from 300 to 800 participants. At the next level are so-called Topical Science Meetings. These sessions focus on more specific topics and attract a somewhat smaller group of participants, typically 80–100 experts in the field. Topical Science Meetings represent a compromise between the very large Open Science Meeting and Working Group Workshops. The latter are smaller, even more specifically oriented meetings usually consisting of 20–40 researchers in an area.

Roger Revelle (1909–1991)

For at least a century after the first discoveries of global warming, scientists were concerned almost entirely with the concentration of carbon dioxide in the air. The studies that were done in the field asked questions such as what the sources and sinks of carbon dioxide were, how the concentration of carbon dioxide in the air had changed over time, what the influence of human activity was on this measure, how increasing levels of carbon dioxide might affect climate, and so on. Almost no one thought of another possible factor in the climate change equation: the oceans. The general opinion, expressed or not, was that the vast quantities of water covering the globe could easily absorb any amount of carbon dioxide emitted to the atmosphere. After all, carbon dioxide is relatively soluble in water, and, the logic went, any additional carbon dioxide added to the atmosphere by human activities could easily be taken up by the oceans.

In fact, scientists knew almost nothing for sure about this theory. As late as the 1950s, for example, no one really knew how long carbon dioxide stayed in the oceans once it was dissolved: a few weeks, a few years, or thousands of years had all be predicted. The answer to that question is of critical importance, of course. If the gas stays dissolved for only a short time, it will escape back into the atmosphere, significantly contributing to the amount present in the atmosphere. If it remained in the oceans for a few hundred or thousand years, atmospheric build-up was much less likely.

In the 1960s, a series of experiments conducted in connection with nuclear weapons testing provided some answers to these questions. As it turned out, the oceans were not the dependable depository for carbon dioxide that had been assumed. As the gas dissolves in seawater, the chemical composition of that water begins to change, reducing its ability to dissolve additional carbon dioxide. The accumulation of carbon dioxide in the atmosphere was, therefore, not a problem to be concerned about for centuries into the future; it posed a potential threat at the present time. And with discoveries such as these, researchers began to realize for the first time that global warming was not a problem to be worried about at some time in the future. Humans had to begin finding ways today to reduce the world's carbon-dioxide problems. In fact, one of those researchers, Roger Revelle, at the time made what has become perhaps the most famous statement about climate change in history. "Human beings," he said, "are now carrying out a large scale geophysical experiment of a kind that could not have happened in the past nor be reproduced in the future." The paper in which this comment appeared, written in conjunction with Hans Suess (see next profile), is also now one of the most famous research reports in the history of climate science (Roger Revelle and Hans Suess. 1957. "Carbon Dioxide Exchange between Atmosphere and Ocean and the Question of an Increase of Atmospheric CO_2 during the Past Decades." *Tellus* 9: 18–27.

https://pdfs.semanticscholar.org/d014/06a57bff758203390e3
6247bd96e0c9f8102.pdf. Accessed on October 2, 2019).

Roger Randall Dougan Revelle was born in Seattle, Washington, on March 7, 1909. He was recognized early in life as a young man of special talents and was admitted to Pomona College at the age of 16. He originally planned to concentrate in journalism but soon turned to geology as his major. After earning his Bachelor's degree in geology from Pomona in 1929, he matriculated at the University of California at Berkeley, from which he earned his PhD in oceanography in 1936. Early in his graduate career, Revelle was recommended by his advisor for a position at the Scripps Institution of Oceanography in San Diego. Scripps was a preeminent center of oceanographic research in the United States at the time (as it continues to be today), and Revelle was assigned to work on some of the ground-breaking studies in the field. Many of these studies were supported by the U.S. government as part of its studies of the physical and chemical properties of the oceans. His historic 1957 paper with Suess grew out of this line of research, in combination with Suess's crucial background in carbon dating.

Upon completion of his doctoral studies, Revelle was offered a position as instructor in oceanography at Scripps. He remained there until the outbreak of World War II, at which time, he joined the U.S. Navy as commander of the oceanographic section of the Bureau of Ships and, later, head of the Navy's geophysics branch. At the war's conclusion, he returned to Scripps, where he eventually served as director from 1951 to 1964.

In 1963, Revelle made a change in his professional career, taking a leave of absence from Scripps to found the Center for Population Studies (CPS) at Harvard University. His primary focus at CPS was the application of science and technology for solving problems of world hunger. In 1976, Revelle left CPS to return to California, where he became professor of science and public policy at the University of California at San Diego (UCSD). He retained his connections with UCSD and Scripps for the rest of his life, offering courses and lectures on a variety of topics, such as problems in Africa, climate change, and marine policy. He died in San Diego on July 15, 1991. Revelle has received a number of honors, including the Alexander Agassiz Medal, Tyler Prize for Environmental Achievement, Vannevar Bush Award, William Bowie Medal, and National Medal of Science. In 1995, Scripps named its new research vessel the R/V Roger Revelle. Four years later the Ocean Studies Board of the National Academies of Sciences, Engineering, and Medicine created the Roger Revelle Commemorative Lecture series. Also, since 1992, the American Geophysical Union has awarded a prize each year in his honor, the Roger Revelle Medal, for outstanding contributions in atmospheric sciences.

Hans Suess (1909–1993)

Hans Eduard Suess was born in Vienna, Austria, on December 16, 1909, into a famous family of geologists. His father was professor of geology and petrography at the University of Vienna, and his grandfather, also a professor at the University of Vienna, president of the Austrian Academy of Sciences, and author of one of the most famous geology textbooks in history, *Das Anlitz der Erde* (The face of the Earth). He attended the University of Vienna, from which he received his PhD in physical chemistry in 1936. The diversity of his interests is reflected in the subject matter of his first dozen papers: heavy water, properties of dissolved styrene, thermal disintegration of dioxane, photochemistry of Earth's atmosphere, radioactivity of potassium, dating the age of meteorites, and the reactions of thermal neutrons.

During the war, Suess became involved with the German war effort, making use especially of his background in the field of heavy water. He was assigned to a plant in Vemork, Norway, where hydrogen gas was being produced from heavy water. He found time on his own to continue his studies on the cosmic abundance of the elements.

In 1949, Suess received an invitation to visit the Institute for Nuclear Studies at the University of Chicago, an offer he accepted a year later. At Chicago, Suess worked in the laboratory of famed chemist Harold Urey, studying the use of the carbon-14 isotope for the dating of ancient objects. It was this background that later became an essential element of the ocean/carbon-dioxide experiments conducted with Roger Revelle (see previous profile) in the early 1950s. In 1951, Suess left Chicago to establish his own carbon dating laboratory at the U.S. Geological Survey in Washington, D.C. Four years later, he accepted an offer from Revelle to join his research team at Scripps. There he created the La Jolla Radiocarbon Laboratory, at which he continued his research on carbon-14 dating. In 1958, Suess became one of the four founding faculty members at the newly created University of California at San Diego. There he held the title of professor of geochemistry. He continued his affiliations with UCSD and Scripps for the rest of his life. He retired officially from UCSD in 1977 and then was name professor emeritus. He died in La Jolla, California, on September 20, 1993.

Greta Thunberg (2003–)

This is all wrong. I shouldn't be up here. I should be back in school on the other side of the ocean. Yet you all come to us young people for hope. How dare you!

You have stolen my dreams and my childhood with your empty words. And yet I'm one of the lucky ones. People are suffering. People are dying. Entire ecosystems are collapsing. We are in the beginning of a mass extinction, and all you can talk about is money and fairy tales of eternal economic growth. How dare you!

("Transcript: Greta Thunberg's Speech at the U.N. Climate Action Summit." 2019. NPR. https://www.npr .org/2019/09/23/763452863/transcript-greta-thunbergs -speech-at-the-u-n-climate-action-summit. Accessed on September 27, 2019)

These were part of the remarks made by Swedish teenager Greta Thunberg at the United Nations Climate Action Summit on September 23, 2019. Thunberg had traveled from her home in Sweden to New York on the racing yacht *Malizia II* because she did not want to contribute to global climate change by flying to her destination. Her speech and trip were only two examples of the message that she had been bringing to the world for almost a decade about the threat posed to the world by climate change.

Greta Tintin Eleonora Ernman Thunberg was born in Stockholm, Sweden, on January 3, 2003. She comes from an artistic tradition: Her mother, Malenda Ernman, is a well-known opera singer, and her father, Svante Thunberg, an actor and author. She is a distant relative of one of the pioneers of climate science Svante Arrhenius, after whom her father was named.

Thunberg's interest in climate change goes back to the third grade, when she first started hearing about the problem in her science classes. She realized the severity of the crisis and wondered why adults were not taking actions to deal with the problem. She began to make changes in her own life, such as becoming a vegan, that she believed would help reduce the risks of climate change. Her family also followed suit, installing a solar power system in their own home and starting a vegetable garden outside their home.

Her battle against climate change was interrupted at the age of 11 when she began to experience physical and mental health problems. She had almost no contact with anyone outside her own family, and she lost so much weight her doctors thought she might starve to death. Eventually her condition was diagnosed as Asperger's syndrome combined with high-functioning autism and obsessive-compulsive disorder. This diagnosis allowed her to begin a medical program that has since allowed her to function as any other teenage girl would be expected to act.

By 2018, however, Thunberg's worries about climate change began to disrupt her life. She started having severe nightmares about the effects of climate change and decided she had to take more aggressive action about the problem. One act she decided on was "Skolstrejk for Klimatet" (School strike for climate). She tried to talk other students and friends into joining her strike for climate change action on the steps of the Swedish Parliament. When she received no support, she just went ahead with the strike on her own. She eventually returned to her post for 21 days, during which she was slowly joined by supporters and, at times, reporters. Her actions had soon won national attention for her cause.

Eventually, Thunberg's climate strikes began to receive attention in other nations, where other students repeated her actions. This movement reached a climax on March 15, 2019, when a global school strike was called, for which an estimated 1.6 million boys and girls in 2,233 cities and 128 countries were said to take part. By some accounts, the action was the largest single climate change protest in history. It was one of the factors that led to her being invited to speak to the United Nations in September 2019.

Thunberg's work has by no means received unanimous support among the media and the general public. As just one example of her critics, radio commentator and former Deputy Assistant to President Donald Trump Sebastian Gorka wrote on his Twitter account that

> This performance by @GretaThunberg is disturbingly redolent of a victim of a Maoist "re-education" camp. The adults who brainwashed this autist child should be brought up on child abuse charges. (Sebastian Gorka DrG 2019. https://twitter.com/SebGorka/status/1176301345151864832. Accessed on September 29, 2019)

In October 2019, the Nordic Council Environmental Prize was awarded to Thunberg; she declined it (and the prize money that goes with it), saying that "the climate movement does not need any more awards" but instead needed people in power to

listen to the science (https://www.cnn.com/2019/10/29/world
/greta-thunberg-nordic-award-decline-trnd/index.html).

350.org

350.org was formed in 2008 by a group of university researchers concerned about the growing problem of climate change.
One of the leaders of the group was American environmental activist, Bill McKibben. McKibben has written 17 books
on the climate change crisis, the first of which was *The End of
Nature*. That book has at times been said to be the first book
on global warming written for the general public. The title of
the organization, 350.org, comes from a commonly expressed
view of researchers that the highest concentration of carbon
dioxide that should exist in the atmosphere is 350 parts per
million (ppm). In fact, that level had already been reached by
the late 1980s. So, the organization's efforts were really directed
at preventing the carbon-dioxide concentration from increasing further and to take such actions as might be necessary to
reduce the level to 350 ppm or less.

One of the organization's first actions was the International
Day of Climate Action in 2009. Concerned groups in countries around the world developed original events to publicize
the threats posed by climate change and actions that could be
taken to deal with the problem. The events included a variety
of displays of the logo "350," by means of individuals laid out
in the shape of the numbers, signs held by divers on coral reefs,
parades of lanterns forming the number, and an aerial piece of
art produced in the Netherlands.

A year later, 350.org sponsored a Global Work Party event,
at which more than 100,000 participants in 176 countries
were said to have organized local events to publicize the problem of climate change and the work of 350.org. The tradition
of engaging local groups in the climate change fight continued in 2011 with the international Moving Planet campaign.
Again, concerned citizens from countries around the world

developed their own ways of publicizing the problem of global warming.

The three basic principles on which 350.org operates are the following:

1. A Fast and Just Transition to 100% Renewable Energy for All
   The best approach to achieving this goal is to support programs developed by local communities.
2. No New Fossil Fuel Projects Anywhere
   Local actions are the most effective way of stopping and banning all coal, oil, and gas projects.
3. Not a Penny More for Dirty Energy
   Such programs should be based on the "3D" approach: divest, desponsor, and defund all public organizations that provide support for fossil fuel companies and activities.

By 2020, 350.org had expanded its membership to more than 300 individual groups worldwide. Examples of those organizations include the Acción Verde Zaragoza, Amazon Watch, Arctic Queen, Bangladesh Youth Movement for Climate, Cairo Bike Scene, Climate Movement of Denmark, Coal Action Network Aotearoa, Costa Rica Neutral, Focus on the Global South, Green Cameroon, Nahdet El Mahrousa's Green Arm, Nepalese Youth for Climate Action, Nigerian Red Cross, Pax Christi USA, Reciclemos Por Panamá, South Pacific Regional Environment Programme, and SustainUS.

In addition to its organizational and coordinating activities, 350.org has developed a variety of resources to be used by its affiliate groups or individuals interested in its activities. Among these resources are videos and animations that illustrate the goals of the organization, as well as records of some of its actions. In addition, it offers a variety of guides, templates, presentations, photos, fonts, and icons that can be used for training and educational purposes.

United Nations Intergovernmental Panel on Climate Change

The Intergovernmental Panel on Climate Change (IPCC) has a long history, dating to an international meeting of climate scientists held in Villach, Austria, in 1985. A major conclusion of the meeting was a consensus that global warming was a reality about which citizens and governments around the world should be better informed. In an effort to deal with that challenge, representatives of the World Meteorological Organization, the United Nations Environment Programme, and the International Committee of Scientific Unions agreed to establish a scientific panel, the Advisory Group on Greenhouse Gases, to work on this issue. Before long, the group realized that the task to which they had been assigned was too large for their size and resources. As a result, largely through the influence of the U.S. government, the United Nations moved to create a larger, better-funded group, the IPCC. The purpose of the new organization was "to prepare a comprehensive review and recommendations with respect to the state of knowledge of the science of climate change; the social and economic impact of climate change, and potential response strategies and elements for inclusion in a possible future international convention on climate" ("History of the IPCC." 2019. IPCC. https://www.ipcc.ch/about/history/. Accessed on October 1, 2019).

IPCC conducts no original research of its own. The mechanism for preparing its reports, instead, begins with an annual meeting of all participants, generally amounting to sessions of hundreds of experts in a variety of fields. This Panel then discusses current issues and selects several specific research topics for the coming year. These topics are then referred to one or more of four standing working groups: The Physical Science Basis; Impacts, Adaptation, and Variability; Mitigation of Climate Change; and National Greenhouse Gas Inventories. The groups then recommend and identify research in each of these areas that can provide information about the current status of each field.

The results of this process are two types of reports. One is an extensive, regular, summary of the state of climate change at some specific point in time. As of 2020, five such reports have been produced: First IPCC Assessment Report (FAR 1990); Second Assessment Report (SAR 1995); Third Assessment Report (TAR 2001); Fourth Assessment Report (AR4 2007); and Fifth Assessment Report (AR5 2014). A Sixth Assessment Report (SAR) is planned for release in 2022. In addition to these general reports, IPCC also issues periodic reports on other specific topics related to climate change. Some subjects covered by those reports include the ocean and the cryosphere, climate change and land, global warming consequences of 1.5°C, renewable energy sources, the ozone layer and climate change, carbon-dioxide capture and storage, and aviation and global atmosphere. A list of and links to all IPCC reports is available on the organization's website at https://www.ipcc.ch/reports/.

United States Environmental Protection Agency

The U.S. Environmental Protection Agency (EPA) was created in 1970 by President Richard M. Nixon as part of his Reorganization Plan No. 3. The action occurred during a period in U.S. history when the nation was becoming aware of a host of serious environmental problems, such as air and water pollution, disposal of toxic wastes, and contamination of the oceans. Among the laws passed during the period were the Clean Air Act of 1963, Water Quality Act of 1965, Solid Waste Disposal Act of 1965, Air Quality Act of 1967, Clean Air Act amendments of 1970, the Federal Water Pollution Control Act amendments of 1972, Coastal Zone Management Act of 1972, and Marine Protection, Research and Sanctuaries Act of 1972. Most important above and beyond these individual acts was the overall legislation contained in the National Environment Act of 1970. The EPA rapidly became a large, well-funded governmental entity for carrying out the provisions of these and many more legislative directives.

Control over climate issues in the EPA came only relatively late in its history. As concerns about global warming developed in the United States, various individuals and groups asked the EPA to begin to monitor carbon-dioxide emissions as a way of controlling climate change. In 2003, the agency announced that it did not believe that it had legislative authority to do so. In response to this action, 12 states and several cities joined together to sue the EPA to reverse its decision and to start regulating carbon-dioxide emissions. That suit slowly worked its way through the judicial system, reaching the U.S Supreme Court in 2006. A year later, that court decided on a 5-to-4 vote in favor of the plaintiffs. It ordered the EPA to begin developing methods for monitoring the release of carbon dioxide as an air pollutant that poses a risk to the environment and human health. (The court case can be found at *Massachusetts v. EPA*, 549 U.S. 497 (2007); https://supreme.justia.com/cases/federal/us/549/497/.)

Under the presidential administration of Barack Obama, the EPA began to enforce climate-change-related laws and regulations more aggressively than had been the case in previous years. For example, it developed a series of rules under which the Clean Power Plan of 2015 were to be used for the control of carbon-dioxide emissions. It also outlined methods by which methane emissions were to be monitored and controlled. The agency also created or more aggressively enforced programs by which transportation and fixed-point sources of greenhouse gas emissions were regulated. The EPA also became an important source of information for the general public about climate change.

The administration of President Donald Trump has taken a somewhat different approach to climate change. Trump has outlined his views on the topic on a number of occasions, usually stating that he thinks that global warming is "a hoax" or, at best, a change that is probably taking place with both good and bad consequences. (For a summary of Trump's views on climate change, see "Donald Trump." 2019. Desmog. https://

www.desmogblog.com/donald-trump. Accessed on October 1, 2019.) In any case, Trump decided to choose one of the most outspoken climate deniers in the country, Scott Pruitt, as his first EPA administrator. He also began to reduce or eliminate EPA's role in the field of climate change early in his administration. Some of the actions taken by the EPA were the removal of the agency's web page devoted specifically to the topic of climate science and to "update" other parts of the agency's website dealing with climate change information. (For details on these changes, see Laignee Barron. 2018. "Here's What the EPA's Website Looks Like after a Year of Climate Change Censorship." *TIME*. https://time.com/5075265/epa-website -climate-change-censorship/. Accessed on October 1, 2019.) How the EPA will deal with the topic of climate change under future presidential administrations cannot, of course, be reliably predicted.

Yale Program on Climate Change Communication

The Yale Program on Climate Change Communication (YPCCC) is part of the university's School of Forestry & Environmental Studies. It was founded as an offshoot of a conference held in Aspen, Colorado, in 2005 on "Americans and Climate Change." More than 100 individuals from fields as diverse as science, media, religion, politics, entertainment, education, business, environmentalism, and civil society met to discuss the risks posed by climate change, current public opinion on the issue, and ways in which those views developed and were influenced by various forces in society. The final report of the conference "Americans and Climate Change: Closing the Gap between Science and Action" contained several recommendations for developing mechanisms by which communication about climate change could be improved. (That report is available online at https://climatecommunication.yale.edu/wp -content/uploads/2016/02/2006_03_Americans-and-Climate

-Change.pdf.) In response to some of those recommendations, YPCCC was created in the fall of 2005.

The focus of YPCCC's activities is the conducting of research on public climate change knowledge, attitudes, policy preferences, and behavior at the global, national, and local scales. The results of that research are disseminated to a wide range of news organizations, such as the television networks ABC, CBS, CNN, NBC, as well as print media, such as the Associated Press and *The Guardian, New York Times,* and *Washington Post* newspapers. Other forms by which research results are disseminated include social media, such as Facebook, LinkedIn, and Twitter; an online publication, "Yale Climate Connections"; and a daily 90-second radio broadcast on climate change.

Current research at YPCCC is directed primarily in six directions. The first, "Audiences," is based on the fact that different groups of people approach the issue of climate change in different ways. YPCCC attempts to identify these groups of people and the way they are influenced by information about climate change. Some of their specific topics in this area have been "Global Warming's Six Americas," "Race, Ethnicity and Public Responses to Climate Change," "Religious Responses to Climate Change," "Generational Responses to Climate Change," and "Five Hurricane Audiences in Coastal Connecticut." A second focus of research has been factors that affect individuals' behaviors and actions with regard to climate change. Examples of the research topics in this area are "Politics & Global Warming, Fall 2015," "The Genesis of Climate Change Activism: From Key Beliefs to Political Action," and "How Americans Communicate about Global Warming." A third line of research deals with the forces that drive people's beliefs and attitudes about climate change. Some studies that have been completed deal with "Americans' Knowledge of Climate Change," "Fracking in the American Mind," and "Climate Change Awareness and Concern in 119 Countries." A fourth category of research, "Climate Change," explores the types of impacts that might be expected from continued climate change. The fifth field of research is called "Messaging." It identifies the best ways by which researchers can explain their research to members of the general public. The sixth area of research, "Policy and Politics," investigates the formal structure through which information about climate change can be provided to policy makers and the general public.

An especially useful tool developed by YPCCC involves the use of visual images for the display of data collected about climate change. As an example, the center releases an annual Yale Climate Opinion Map, which illustrates the ways in which people's understandings of and beliefs about climate change vary from state to state in the United States. Another visualization tool is the "Six Americas Survey Super Short Survey (SASSY)," which summarizes views of six categories of Americans about basic features of climate change. The six categories represented in the surveys are those who are "alarmed," "concerned," "cautious," "disengaged," "doubtful," and "dismissive." A complete list of all visualization and data reports released by YPCCC can be found at https://climatecommunication.yale .edu/visualizations-data/.

One way to better understanding the debate about climate change is to examine data and documents related to the issue. The data here provides indicators of climate change, while the documents review the history and current events on the subject.

Data

This table shows the amount of radiative forcing caused by five different substances from 1979 to 2018.

_____

A drought-stricken marsh exemplifies the extreme effects of global climate change. (Tom Linster/Dreamstime.com)

Table 5.1 Global Radiative Forcing, CO_2-equivalent Mixing Ratio, and the AGGI (1979–2018)*

| | Global Radiative Forcing (W/m^{-2}) | | | | | | | CO_2-eq (ppm) |
Year	CO_2	CH_4	N_2O	CFC12	CFC11	15-minor	Total	Total
1979	1.027	0.406	0.104	0.092	0.040	0.031	1.699	382
1980	1.058	0.413	0.104	0.097	0.042	0.034	1.748	385
1981	1.077	0.420	0.107	0.102	0.044	0.036	1.786	388
1982	1.089	0.426	0.111	0.107	0.046	0.038	1.818	391
1983	1.115	0.429	0.113	0.113	0.048	0.041	1.859	394
1984	1.140	0.432	0.116	0.118	0.050	0.044	1.900	397
1985	1.162	0.437	0.118	0.123	0.053	0.047	1.940	400
1986	1.184	0.442	0.121	0.129	0.056	0.049	1.982	403
1987	1.211	0.447	0.120	0.136	0.058	0.053	2.025	406
1988	1.250	0.451	0.122	0.143	0.061	0.057	2.085	410
1989	1.275	0.455	0.126	0.149	0.063	0.061	2.130	414
1990	1.293	0.459	0.129	0.154	0.065	0.065	2.165	417
1991	1.312	0.463	0.131	0.158	0.066	0.069	2.199	419
1992	1.323	0.467	0.133	0.162	0.067	0.072	2.224	421
1993	1.334	0.467	0.133	0.164	0.067	0.074	2.239	422
1994	1.356	0.470	0.135	0.165	0.067	0.076	2.269	425
1995	1.383	0.472	0.136	0.168	0.067	0.077	2.303	428
1996	1.410	0.473	0.139	0.170	0.066	0.078	2.335	430
1997	1.426	0.474	0.142	0.171	0.066	0.079	2.357	432
1998	1.464	0.478	0.144	0.172	0.066	0.080	2.404	436

Year								
1999	1.495	0.481	0.148	0.173	0.065	0.082	2.443	439
2000	1.513	0.481	0.151	0.173	0.065	0.083	2.466	441
2001	1.535	0.480	0.153	0.174	0.064	0.085	2.492	443
2002	1.564	0.481	0.155	0.174	0.064	0.087	2.525	446
2003	1.600	0.483	0.157	0.174	0.063	0.088	2.566	449
2004	1.627	0.483	0.159	0.174	0.063	0.090	2.596	452
2005	1.655	0.482	0.162	0.173	0.062	0.092	2.626	454
2006	1.685	0.482	0.165	0.173	0.062	0.095	2.661	457
2007	1.710	0.484	0.167	0.172	0.061	0.098	2.692	460
2008	1.739	0.486	0.170	0.171	0.061	0.100	2.728	463
2009	1.760	0.489	0.172	0.171	0.060	0.103	2.755	465
2010	1.791	0.491	0.175	0.169	0.060	0.106	2.792	468
2011	1.817	0.492	0.178	0.169	0.059	0.109	2.824	471
2012	1.845	0.494	0.181	0.168	0.059	0.112	2.858	474
2013	1.882	0.496	0.183	0.167	0.058	0.114	2.900	478
2014	1.908	0.499	0.187	0.166	0.058	0.117	2.935	481
2015	1.938	0.504	0.190	0.165	0.058	0.119	2.974	485
2016	1.985	0.507	0.193	0.164	0.057	0.122	3.028	490
2017	2.013	0.509	0.195	0.163	0.057	0.124	3.062	493
2018	2.044	0.512	0.199	0.162	0.057	0.127	3.101	496

*Radiative forcing is defined as the difference between incoming and outgoing radiation as the result of some given factor (in this case, greenhouse gases).

Source: Butler, James H., and Stephen A. Montzka. 2019. "The NOAA Annual Greenhouse Gas Index (AGGI)." Global Monitoring Division. Earth System Research Laboratory. https://www.esrl.noaa.gov/gmd/aggi/aggi.html. Accessed on August 12, 2019.

This table shows the annual data on which the Keeling curve is based, that is, the concentration of carbon dioxide in the atmosphere about Mauna Loa, Hawaii, from 1959 to 2018.

Table 5.2   Average Annual CO_2 Concentration, Mauna Loa, Hawaii parts per million (ppm; uncertainty in all measurements is 0.12 ppm)

Year	Concentration
1959	315.97
1960	316.91
1961	317.64
1962	318.45
1963	318.99
1964	319.62
1965	320.04
1966	321.38
1967	322.16
1968	323.04
1969	324.62
1970	325.68
1971	326.32
1972	327.45
1973	329.68
1974	330.18
1975	331.11
1976	332.04
1977	333.83
1978	335.40
1979	336.84
1980	338.75
1981	340.11
1982	341.45
1983	343.05
1984	344.65
1985	346.12
1986	347.42
1987	349.19
1988	351.57
1989	353.12
1990	354.39
1991	355.61

(continued)

Table 5.2 *(continued)*

Year	Concentration
1992	356.45
1993	357.10
1994	358.83
1995	360.82
1996	362.61
1997	363.73
1998	366.70
1999	368.38
2000	369.55
2001	371.14
2002	373.28
2003	375.80
2004	377.52
2005	379.80
2006	381.90
2007	383.79
2008	385.60
2009	387.43
2010	389.90
2011	391.65
2012	393.85
2013	396.52
2014	398.65
2015	400.83
2016	404.24
2017	406.55
2018	408.52

Source: "Mauna Loa CO_2 Annual Mean Data." 2019. Trends in Atmospheric Carbon Dioxide. Global Monitoring Division. Earth System Research Laboratory. https://www.esrl.noaa.gov/gmd/ccgg/trends/data.html. Accessed on August 14, 2019.

This table summarizes changes in ice covering of the Arctic Sea from 1979 to 2017.

Table 5.3   Area Covered by Ice in Arctic Sea, 1979–2017 (million square miles)

Year	Area
1979	6.455322
1980	7.010784
1981	6.438129
1982	6.662314
1983	6.785270
1984	6.206512
1985	6.140910
1986	6.663552
1987	6.248912
1988	6.492643
1989	6.265363
1990	5.635686
1991	5.733746
1992	6.499373
1993	5.469105
1994	6.273856
1995	5.544039
1996	6.752524
1997	6.133342
1998	5.607585
1999	5.200637
2000	5.512233
2001	5.879375
2002	5.126711
2003	5.393702
2004	5.399485
2005	4.951513
2006	5.160366
2007	3.712601
2008	4.033307
2009	4.587511
2010	4.106893
2011	3.734480
2012	2.910646

(continued)

Table 5.3 (*continued*)

Year	Area
2013	4.645969
2014	4.527100
2015	3.813979
2016	3.560391
2017	4.172772

Source: "Arctic Sea Ice Minimum." 2019. Global Climate Change. NASA. https://climate.nasa.gov/vitalsigns/arcticseaice/. Accessed on August 14, 2019.

This table summarizes the concentration (in parts per million) of carbon dioxide in Earth's atmosphere over just a century.

Table 5.4   Atmospheric CO_2 Levels, 1850–1958 (parts per million)

1850	285.2
1851	285.1
1852	285.0
1853	285.0
1854	284.9
1855	285.1
1856	285.4
1857	285.6
1858	285.9
1859	286.1
1860	286.4
1861	286.6
1862	286.7
1863	286.8
1864	286.9
1865	287.1
1866	287.2
1867	287.3
1868	287.4
1869	287.5
1870	287.7
1871	287.9
1872	288.0
1873	288.2

(*continued*)

Table 5.4 (*continued*)

1874	288.4
1875	288.6
1876	288.7
1877	288.9
1878	289.5
1879	290.1
1880	290.8
1881	291.4
1882	292.0
1883	292.5
1884	292.9
1885	293.3
1886	293.8
1887	294.0
1888	294.1
1889	294.2
1890	294.4
1891	294.6
1892	294.8
1893	294.7
1894	294.8
1895	294.8
1896	294.9
1897	294.9
1898	294.9
1899	295.3
1900	295.7
1901	296.2
1902	296.6
1903	297.0
1904	297.5
1905	298.0
1906	298.4
1907	298.8
1908	299.3
1909	299.7
1910	300.1
1911	300.6
1912	301.0
1913	301.3

(*continued*)

Table 5.4 *(continued)*

1914	301.4
1915	301.6
1916	302.0
1917	302.4
1918	302.8
1919	303.0
1920	303.4
1921	303.7
1922	304.1
1923	304.5
1924	304.9
1925	305.3
1926	305.8
1927	306.2
1928	306.6
1929	307.2
1930	307.5
1931	308.0
1932	308.3
1933	308.9
1934	309.3
1935	309.7
1936	310.1
1937	310.6
1938	311.0
1939	311.2
1940	311.3
1941	311.0
1942	310.7
1943	310.5
1944	310.2
1945	310.3
1946	310.3
1947	310.4
1948	310.5
1949	310.9
1950	311.3
1951	311.8
1952	312.2
1953	312.6

(continued)

Table 5.4 *(continued)*

1954	313.2
1955	313.7
1956	314.3
1957	314.8
1958	315.3

Note: For data from 1959 to the present, see Table 5.2.
Source: "Global Mean CO₂ Mixing Ratios (ppm): Observations." 2015. Goddard Institute for Space Studies. National Aeronautics and Space Administration.   https://data.giss.nasa.gov/modelforce/ghgases/Fig1A.ext.txt. Accessed on August 15, 2019.

Documents

On the Influence of Carbonic Acid in the Air upon the Temperature of the Ground (1896)

This classic paper by Swedish chemist and physicist Svante Arrhenius, for the first time in history, developed a clear and convincing relationship between the amount of carbon dioxide in the air and Earth's temperature. From this long, detailed, and technical report, the following sections contain the most important conclusions arising from his work. Note that carbon dioxide is referred to, as was common at the time, as "carbonic acid," the aqueous form of carbon dioxide [H_2CO_3].

I should certainly not have undertaken these tedious calculations if an extraordinary interest had not been connected with them. In the Physical Society of Stockholm there have been occasionally very lively discussions on the probable causes of the Ice Age; and these discussions have, in my opinion, led to the conclusion that there exists as yet no satisfactory hypothesis that could explain how the climatic conditions for an ice age could be realized in so short a time as that which has elapsed from the days of the glacial epoch.

[Arrhenius next reviews existing information about the occurrence of ice ages in various parts of the world in geologic history.]

. . .

One may now ask, How much must the carbonic acid vary according to our figures, in order that the temperature should attain the same values as in the Tertiary and Ice ages respectively? A simple calculation shows that the temperature in the arctic regions would rise about 8° to 9°C, if the carbonic acid increased to 2.5 or 3 times its present value. In order to get the temperature of the ice age between the 40th and 50th parallels, the carbonic acid in the air should sink to 0.62–0.55 of its present value (lowering of temperature 4°–5°C.). The demands of the geologists, that at the genial epochs the climate should be more uniform than now, accords very well with our theory. The geographical annual and diurnal ranges of temperature would be partly smoothed away, if the quantity of carbonic acid was augmented. The reverse would be the case (at least to a latitude of 50° from the equator) if the carbonic acid diminished in amount. But in both these cases I incline to think that the secondary action (see p. 257) due to the regress or the progress of the snow-covering would play the most important role.

. . .

There is now an important question which should be answered, namely:—Is it probable that such great variations in the quantity of carbonic acid as our theory requires have occurred in relatively short geological times? The answer to this question is given by Prof. Högbom. As his memoir on this question may not be accessible to most readers of these pages, I have summed up and translated his utterances which are of most importance to our subject *:—

[Arrhenius then provides a lengthy quote by Högbom on the sources of carbon dioxide in the atmosphere and variations in the contribution of each source over time. Högbom's conclusion on the question is as follows:]

"If we pass the above-mentioned processes for consuming and producing carbonic acid under review, we find that they evidently do not stand in such a relation to or dependence on one another that any probability exists for the permanence of an equilibrium of the carbonic acid in the atmosphere. An increase or decrease of the supply continued during geological periods must, although it may not be important, conduce to remarkable alterations of the quantity of carbonic acid in the air, and there is no conceivable hindrance to imagining that this might in a certain geological period have been several times greater, or on the other hand considerably less, than now."

[Arrhenius concludes, then, that]

"the question of the probability of quantitative variation of the carbonic acid in the atmosphere is in the most decided manner answered by Prof. Högbom . . ."

Source: Arrhenius, Svante. 1896. "On the Influence of Carbonic Acid in the Air upon the Temperature of the Ground." *The London, Edinburgh, and Dublin Philosophical Magazine and Journal of Science* 41(251): 237–276. doi: 10.1080/14786449608620846. http://empslocal.ex.ac.uk /people/staff/gv219/classics.d/Arrhenius96.pdf. Accessed on August 12, 2019.

Remarkable Weather of 1911 (1912)

Arrhenius's research on the connection between carbon dioxide in the atmosphere and climate was of interest to academicians in the area. But it also had somewhat subtle influences on the general public, primarily through popular journals and magazines that picked up on his ideas and pursued their possible effects on Earth's future climate. This segment comes from one such article.

It has been found that if the air contained more carbon dioxide, which is the product of the combustion of coal or vegetable material, the temperature would be somewhat higher. In fact, a theory has been elaborated by the great Swedish scientist Arrhenius, that the earth has had a warm climate when the amount of carbon dioxide in the air was abundant, and a cold climate when it was scarce. It is believed that if the atmosphere contained two to three times its present amount, the climate would be considerably warmer, and if it should lose half of that which it now has, the glaciers would again form in Canada. There are good reasons for believing that the quantity of this gas in the atmosphere may slowly undergo variations.

Since burning coal produces carbon dioxide it may be inquired whether the enormous use of that fuel in modern times may not be an important factor in filling the atmosphere with this substance, and consequently in indirectly raising the temperature of the earth. In the United States about 500,000,000 tons of coal were mined in 1911. Suppose four times this amount were mined and burned in the whole world. When this amount is burned, 7,000,000,000 tons of carbon dioxide are put into the atmosphere. The question is, simply, whether this is an appreciable fraction of that which the atmosphere already holds, and whether there are any important ways in which it is being removed from the atmosphere.

The atmosphere contains 1,500,000,000,000 tons of carbon dioxide. Consequently the combustion of coal at the present rate will double it in about 200 years, unless it is removed by some means in enormous quantities. Carbon dioxide is removed from the atmosphere by growing plants and in fact the carbon in coal came from the air through the vegetable matter from which it has been formed. But when vegetable matter is burned, or decays, or is consumed by animals, the carbon dioxide is returned to the atmosphere. It does not seem that there will be any great gain or loss in the next few centuries. A more important factor is the oceans which now hold enormous quantities of carbon dioxide and which, under suitable conditions, can absorb much more. In fact, they are the great regulators and have been involved essentially in all the variations of the past. But the action of the sea is very slow, and it may well be that the enormous present-day combustion of coal is producing carbon dioxide so fast that it will have important climatic effects.

Source: Molena, Francis. 1912. "Remarkable Weather of 1912." *Popular Mechanics*, March 1912: 339–342. https://ia601307 .us.archive.org/10/items/PopularMechanics1912/Popular_Me chanics_03_1912.pdf. Accessed on August 13, 2019.

Energy Tax Act (1978)

For more than four decades, the U.S. Congress has accepted the concept of providing tax incentives for the development and use of alternative forms of energy, such as biomass, nuclear, solar, and wind energy. It has passed numerous acts and amendments to act to achieve this object. At first, this concept was developed as a way of reducing the nation's dependence on foreign gas and oil imports. Over the years, such legislation has also been popular as a way of weaning the country away from the use of fossil fuels, and, hence, from the release of carbon dioxide into the atmosphere. The precise form of these acts has varied from session to session, but this early version of the legislation provides a simple overview of the type of incentives thought to be effective by the Congress. A useful guide to the history of legislation in the United States providing tax incentives for renewable energy is available at "Climate Change Laws in the USA." 2018. Climate Home News. https:// www.climatechangenews.com/2013/02/12/in-focus-usas-climate -laws/. Accessed on October 2, 2019. The main features of the 1978 act are as follows.

Summary: H.R.5263—95th Congress (1977–1978)

. . .

Title I: Residential Energy Credit—Amends the Internal Revenue Code to allow an income tax credit to an individual for an amount equal to the sum of: (1) 15 percent of the energy conservation expenditures up to a maximum of $2,000; and (2) 30 percent of qualified renewable energy source expenditures for solar, wind, and geothermal energy equipment as does not exceed $2,000 plus 20 percent of such expenditures as exceeds $2,000 but does not exceed $10,000. Provides for a credit carryover to the extent that such credit exceeds the taxpayer's tax liability.

Title II: Transportation—Imposes a gas guzzler excise tax on the sale by a manufacturer of each automobile that falls below a specified gasoline efficiency standard for each model year. Imposes such tax on automobiles weighing less than 6,000 pounds. Imposes a tax equal to $200 for an automobile efficiency rating of 14 to 15 miles per gallon in 1980 which is increased to $1,800 in 1985. Imposes such tax at a maximum rate equal to $550 in 1980 for efficiency ratings below 13 miles per gallon and increased to $3,850 in 1986 for efficiency ratings of less than 12.5 miles per gallon. Exempts certain vehicles from the tax.

. . .

Title III: Changes in Business Investment Credit to Encourage Conservation of, or Conversion from, Oil and Gas or to Encourage New Energy Technology—Allows a ten percent tax credit to offset tax liability for each of the following qualified investments placed in service between October 1, 1978, and January 1, 1983: (1) alternative energy property such as nuclear, geothermal, and solar power equipment; (2) specially defined energy property which reduces the amount of energy consumed; and (3) energy property such as recycling equipment, shale oil equipment, and equipment for producing natural gas from geopressured brine.

Title IV: Miscellaneous Provisions—Provides for a depletion deduction for geothermal resources located in the United States beginning at 22 percent on October 1, 1978, and decreasing to 15 percent after 1983. Provides for a ten percent depletion deduction for natural gas produced from geopressured brine from wells drilled from October 1, 1978, to December 30, 1983.

Grants the option to deduct as expenses intangible drilling and development costs in the case of wells drilled in the United States, for any geothermal deposit.

Continues and extends the minimum tax provisions for excess intangible drilling costs of individuals which are applicable to oil and gas production income to include income from geothermal resources and natural gas produced from geopressured brine. Extends the recapture provisions for oil and gas property to include geothermal wells. Extends the risk limitation of losses which may be deducted for exploiting oil and gas to include geothermal resources.

Source: Summary: H.R.5263—95th Congress (1977–1978). 1978. Congress.gov. https://www.congress.gov/bill/95th-congress/house-bill/5263/summary/48. Accessed on October 2, 2019.

Massachusetts v. EPA (2007)

In 2006, the state of Massachusetts, 11 other states, several cities, and a number of other organizations sought to require the Environmental Protection Agency to regulate carbon dioxide and other greenhouse gases (GHG) under its authority to regulate pollutants. They pointed out the Clean Air Act of 1963 allowed the EPA to regulate "pollutants," and that GHG had now clearly been shown to fit into that category. They posed two questions to the court about the EPA's lack of action (as in the selection below). In response, 10 states and 6 trade associations raised the issue of standing with the complainants. That is, they said that none of the 12 states, cities, or other organizations would themselves suffer harm from the EPA's decision not to regulate GHG. In a 5-to-4 vote, the court sided with the complainants and said that the EPA could regulate GHG as air pollutants. This selection is from the syllabus. All reference to text of the decision, notes, and citations have been omitted, as indicated with triple asterisks (***).

Held:
1. *Petitioners have standing to challenge the EPA's denial of their rulemaking petition. Pp. 12–23.*

(a) *This case suffers from none of the defects that would preclude it from being a justiciable Article III "Controvers[y]."*

. . .

(b) The harms associated with climate change are serious and well recognized. The Government's own objective assessment of the relevant science and a strong consensus among qualified experts indicate that global warming threatens, *inter alia*, a precipitate rise in sea levels, severe and irreversible changes to natural ecosystems, a significant reduction in winter snowpack with direct and important economic consequences, and increases in the spread of disease and the ferocity of weather events. That these changes are widely shared does not minimize Massachusetts' interest in the outcome of this litigation. *** According to petitioners' uncontested affidavits, global sea levels rose between 10 and 20 centimeters over the 20th century as a result of global warming and have already begun to swallow Massachusetts' coastal land. Remediation costs alone, moreover, could reach hundreds of millions of dollars. ***

(c) Given EPA's failure to dispute the existence of a causal connection between man-made greenhouse gas emissions and global warming, its refusal to regulate such emissions, at a minimum, "contributes" to Massachusetts' injuries. EPA overstates its case in arguing that its decision not to regulate contributes so insignificantly to petitioners' injuries that it cannot be hauled into federal court, and that there is no realistic possibility that the relief sought would mitigate global climate change and remedy petitioners' injuries, especially since predicted increases in emissions from China, India, and other developing nations will likely offset any marginal domestic decrease EPA regulation could bring about. Agencies, like legislatures, do not generally resolve massive problems in one fell swoop, *** but instead whittle away over time, refining their approach as circumstances change and they develop a more nuanced understanding of how best to proceed, *** That a first step might be tentative does not by itself negate federal-court jurisdiction. And reducing domestic automobile emissions is hardly tentative. Leaving aside the other greenhouse gases, the record indicates that the U.S. transportation sector emits an enormous quantity of carbon dioxide into the atmosphere.***

(d) While regulating motor-vehicle emissions may not by itself reverse global warming, it does not follow that the Court lacks jurisdiction to decide whether EPA has a duty to take steps to slow or reduce it. *** Because of the enormous potential consequences, the fact that a remedy's effectiveness might be delayed during the (relatively short) time it takes for a new motor-vehicle fleet to replace an older one is essentially irrelevant. Nor is it dispositive that developing countries are poised to substantially increase greenhouse gas emissions: A reduction in domestic emissions would slow the pace of global emissions increases, no matter what happens elsewhere. The Court attaches considerable significance to EPA's espoused belief that global climate change must be addressed. ***

. . .

4. EPA's alternative basis for its decision—that even if it has statutory authority to regulate greenhouse gases, it would be unwise to do so at this time—rests on reasoning divorced from the statutory text. While the statute conditions EPA action on its formation of a "judgment," that judgment must relate to whether an air pollutant "cause[s], or contribute[s] to, air pollution which may reasonably be anticipated to endanger public health or welfare." *** Under the Act's clear terms, EPA can avoid promulgating regulations only if it determines that greenhouse gases do not contribute to climate change or if it provides some reasonable explanation as to why it cannot or

will not exercise its discretion to determine whether they do. It has refused to do so, offering instead a laundry list of reasons not to regulate, including the existence of voluntary Executive Branch programs providing a response to global warming and impairment of the President's ability to negotiate with developing nations to reduce emissions. These policy judgments have nothing to do with whether greenhouse gas emissions contribute to climate change and do not amount to a reasoned justification for declining to form a scientific judgment. Nor can EPA avoid its statutory obligation by noting the uncertainty surrounding various features of climate change and concluding that it would therefore be better not to regulate at this time. If the scientific uncertainty is so profound that it precludes EPA from making a reasoned judgment, it must say so. The statutory question is whether sufficient information exists for it to make an endangerment finding. Instead, EPA rejected the rulemaking petition based on impermissible considerations. Its action was therefore "arbitrary, capricious, or otherwise not in accordance with law," *** On remand, EPA must ground its reasons for action or inaction in the statute. ***.

415 F. 3d 50, reversed and remanded.

Source: *Massachusetts v. EPA*, 549 U.S. 497 (2007).

Executive Order 13514 (2009)

The administration of President Barack Obama (2008–2017) was especially active in putting forward programs to make the American public aware of climate change issues and developing policies and practices to deal with that problem. One of the president's first major actions in this area was the release of a long Executive Order on October 5, 2009. The essence of that document is expressed in the following sections. The document begins with a statement of federal policy, followed by specific actions that are to be taken by government agencies. This Executive Order was rescinded by President Donald Trump's Executive Order 13783, of March 28, 2017.

Section 1. Policy. In order to create a clean energy economy that will increase our Nation's prosperity, promote energy security, protect the interests of taxpayers, and safeguard the health of our environment, the Federal Government must lead by example. It is therefore the policy of the United States that Federal agencies shall increase energy efficiency; measure, report, and reduce their greenhouse gas emissions from direct and indirect activities; conserve and protect water resources through efficiency, reuse, and stormwater management; eliminate waste, recycle, and prevent pollution; leverage agency acquisitions to foster markets for sustainable technologies and environmentally preferable materials, products, and services; design, construct, maintain, and operate high performance sustainable buildings in sustainable locations; strengthen the vitality and livability of the communities in which Federal facilities are located; and inform Federal employees about and involve them in the achievement of these goals.

It is further the policy of the United States that to achieve these goals and support their respective missions, agencies shall prioritize actions based on a full accounting of both economic and social benefits and costs and shall drive continuous improvement by annually evaluating performance, extending or expanding projects that have net benefits, and reassessing or discontinuing under-performing projects.

Finally, it is also the policy of the United States that agencies' efforts and outcomes in implementing this order shall be transparent and that agencies shall therefore disclose results associated with the actions taken pursuant to this order on publicly available Federal websites.

Sec. 2. Goals for Agencies. In implementing the policy set forth in section 1 of this order, and preparing and implementing the Strategic Sustainability Performance Plan called for in section 8 of this order, the head of each agency shall:

(a) within 90 days of the date of this order, establish and report to the Chair of the Council on Environmental Quality (CEQ Chair) and the Director of the Office of Management and Budget (OMB Director) a percentage reduction target for agency-wide reductions of scope 1 and 2 greenhouse gas emissions in absolute terms by fiscal year 2020, relative to a fiscal year 2008 baseline of the agency's scope 1 and 2 greenhouse gas emissions.

[Each of the following lettered sections is followed by specific actions to be taken to achieve the stated goal.]

. . .

(b) within 240 days of the date of this order and concurrent with submission of the Strategic Sustainability Performance Plan as described in section 8 of this order, establish and report to the CEQ Chair and the OMB Director a percentage reduction target for reducing agency-wide scope 3 greenhouse gas emissions in absolute terms by fiscal year 2020, relative to a fiscal year 2008 baseline of agency scope 3 emissions.

. . .

(c) establish and report to the CEQ Chair and OMB Director a comprehensive inventory of absolute greenhouse gas emissions, including scope 1, scope 2, and specified scope 3 emissions . . .

. . .

(d) improve water use efficiency and management . . .

. . .

(e) promote pollution prevention and eliminate waste . . .

. . .

(f) advance regional and local integrated planning . . .

. . .

(g) implement high performance sustainable Federal building design, construction, operation and management,
maintenance, and deconstruction . . .

. . .

(h) advance sustainable acquisition to ensure that 95 percent of new contract actions including task and delivery
orders, for products and services with the exception of
acquisition of weapon systems, are energy efficient . . .

. . .

(i) promote electronics stewardship . . .

. . .

(j) sustain environmental management . . .

Source: "Federal Leadership in Environmental, Energy, and
Economic Performance." 2009. Presidential Documents. *Federal Register* 74(194): 52117–52127. https://www.govinfo.gov
/content/pkg/FR-2009-10-08/pdf/E9-24518.pdf. Accessed on
August 13, 2019.

President Obama's Climate Action Plan (2013)

*In June 2013, President Barack Obama announced a broad-based
plan that would reduce the release of carbon into the atmosphere,
thus cutting back on the climate change the nation and the world
are currently experiencing. The following summary highlights the
main features of the plan. The plan was rescinded in President Donald Trump's Executive Order 13783 of March 28, 2017 (see below).*

President Obama's Plan to Cut Carbon Pollution

Taking Action for Our Kids

We have a moral obligation to leave our children a planet that's
not polluted or damaged, and by taking an all-of-the-above

approach to develop homegrown energy and steady, responsible steps to cut carbon pollution, we can protect our kids' health and begin to slow the effects of climate change so we leave a cleaner, more stable environment for future generations. Building on efforts underway in states and communities across the country, the President's plan cuts carbon pollution that causes climate change and threatens public health. Today, we have limits in place for arsenic, mercury, and lead, but we let power plants release as much carbon pollution as they want—pollution that is contributing to higher rates of asthma attacks and more frequent and severe floods and heat waves.

Cutting carbon pollution will help keep our air and water clean and protect our kids. The President's plan will also spark innovation across a wide variety of energy technologies, resulting in cleaner forms of American-made energy and cutting our dependence on foreign oil. Combined with the President's other actions to increase the efficiency of our cars and household appliances, the President's plan will help American families cut energy waste, lowering their gas and utility bills. In addition, the plan steps up our global efforts to lead on climate change and invests to strengthen our roads, bridges, and shorelines so we can better protect people's homes, businesses, and way of life from severe weather.

While no single step can reverse the effects of climate change, we have a moral obligation to act on behalf of future generations. Climate change represents one of the major challenges of the 21st century, but as a nation of innovators, we can and will meet this challenge in a way that advances our economy, our environment, and public health all at the same time. That is why the President's comprehensive plan takes action to:

Cuts Carbon Pollution in America. In 2012, U.S. carbon pollution from the energy sector fell to the lowest level in two decades even as the economy continued to grow. To build on this progress, the Obama Administration is putting in place tough new rules to cut carbon pollution—just like we have for other toxins like mercury and arsenic—so we protect the health of our children and move our economy toward American-made clean energy sources that will create good jobs and lower home energy bills. For example, the plan:

- Directs EPA to work closely with states, industry and other stakeholder to establish carbon pollution standards for both new and existing power plants;

- Makes up to $8 billion in loan guarantee authority available for a wide array of advanced fossil energy and efficiency projects to support investments in innovative technologies;

- Directs DOI to permit enough renewables project—like wind and solar—on public lands by 2020 to power more than 6 million homes; designates the first-ever hydropower project for priority permitting; and sets a new goal to install 100 megawatts of renewables on federally assisted housing by 2020; while maintaining the commitment to deploy renewables on military installations;

- Expands the President's Better Building Challenge, focusing on helping commercial, industrial, and multi-family buildings cut waste and become at least 20 percent more energy efficient by 2020;

- Sets a goal to reduce carbon pollution by at least 3 billion metric tons cumulatively by 2030—more than half of the annual carbon pollution from the U.S. energy sector—through efficiency standards set over the course of the Administration for appliances and federal buildings;

- Commits to partnering with industry and stakeholders to develop fuel economy standards for heavy-duty vehicles to save families money at the pump and further reduce reliance on foreign oil and fuel consumption post-2018; and

- Leverages new opportunities to reduce pollution of highly-potent greenhouse gases known as hydrofluorocarbons;

directs agencies to develop a comprehensive methane strategy; and commits to protect our forests and critical landscapes.

Prepares the United States for the Impacts of Climate Change. Even as we take new steps to cut carbon pollution, we must also prepare for the impacts of a changing climate that are already being felt across the country. Building on progress over the last four years, the plan:

• Directs agencies to support local climate-resilient investment by removing barriers or counterproductive policies and modernizing programs; and establishes a short-term task force of state, local, and tribal officials to advise on key actions the Federal government can take to help strengthen communities on the ground;

• Pilots innovative strategies in the Hurricane Sandy-affected region to strengthen communities against future extreme weather and other climate impacts; and building on a new, consistent flood risk reduction standard established for the Sandy-affected region, agencies will update flood-risk reduction standards for all federally funded projects;

• Launches an effort to create sustainable and resilient hospitals in the face of climate change through a public-private partnership with the healthcare industry;

• Maintains agricultural productivity by delivering tailored, science-based knowledge to farmers, ranchers, and landowners; and helps communities prepare for drought and wildfire by launching a National Drought Resilience Partnership and by expanding and prioritizing forest- and rangeland- restoration efforts to make areas less vulnerable to catastrophic fire; and

• Provides climate preparedness tools and information needed by state, local, and private-sector leaders through a centralized "toolkit" and a new Climate Data Initiative.

Lead International Efforts to Address Global Climate Change. Just as no country is immune from the impacts of climate change, no country can meet this challenge alone. That is why it is imperative for the United States to couple action at home with leadership internationally. America must help forge a truly global solution to this global challenge by galvanizing international action to significantly reduce emissions, prepare for climate impacts, and drive progress through the international negotiations. For example, the plan:

• Commits to expand major new and existing international initiatives, including bilateral initiatives with China, India, and other major emitting countries;

• Leads global sector public financing towards cleaner energy by calling for the end of U.S. government support for public financing of new coal-fired powers plants overseas, except for the most efficient coal technology available in the world's poorest countries, or facilities deploying carbon capture and sequestration technologies; and

• Strengthens global resilience to climate change by expanding government and local community planning and response capacities.

Source: "President Obama's Plan to Cut Carbon Pollution." 2013. The White House. https://obamawhitehouse.archives .gov/the-press-office/2013/06/25/fact-sheet-president-obama -s-climate-action-plan. Accessed on August 13, 2019. Full report is available at https://obamawhitehouse.archives.gov/sites /default/files/image/president27sclimateactionplan.pdf.

Executive Order 13653 (2013)

The second of President Obama's two main Executive Orders dealing with climate change was released on November 1, 2013. That order restated his administration's position on the importance of climate change in the United States and the world and outlined a set of policies to be put into place in the federal government. Only brief summaries of the main sections are provided here.

Section 1. Policy. The impacts of climate change—including an increase in prolonged periods of excessively high temperatures, more heavy downpours, an increase in wildfires, more severe droughts, permafrost thawing, ocean acidification, and sea-level rise—are already affecting communities, natural resources, ecosystems, economies, and public health across the Nation. These impacts are often most significant for communities that already face economic or health-related challenges, and for species and habitats that are already facing other pressures. Managing these risks requires deliberate preparation, close cooperation, and coordinated planning by the Federal Government, as well as by stakeholders, to facilitate Federal, State, local, tribal, private-sector, and nonprofit-sector efforts to improve climate preparedness and resilience; help safeguard our economy, infrastructure, environment, and natural resources; and provide for the continuity of executive department and agency (agency) operations, services, and programs.

. . .

The Federal Government must build on recent progress and pursue new strategies to improve the Nation's preparedness and resilience. In doing so, agencies should promote: (1) engaged and strong partnerships and information sharing at all levels of government; (2) risk-informed decisionmaking and the tools to facilitate it; (3) adaptive learning, in which experiences serve as opportunities to inform and adjust future actions; and (4) preparedness planning.

Sec. 2. Modernizing Federal Programs to Support Climate Resilient Investment. (a) To support the efforts of regions, States, local communities, and tribes, all agencies, consistent with their missions and in coordination with the Council on Climate Preparedness and Resilience (Council) established in section 6 of this order, shall:

(i)   identify and seek to remove or reform barriers that discourage investments or other actions to increase the Nation's resilience to climate change while ensuring continued protection of public health and the environment;

(ii) reform policies and Federal funding programs that may, perhaps unintentionally, increase the vulnerability of natural or built systems, economic sectors, natural resources, or communities to climate change related risks;

(iii) identify opportunities to support and encourage smarter, more climate-resilient investments by States, local communities, and tribes, including by providing incentives through agency guidance, grants, technical assistance, performance measures, safety considerations, and other programs, including in the context of infrastructure development as reflected in Executive Order 12893 of January 26, 1994 (Principles for Federal Infrastructure Investments), my memorandum of August 31, 2011 (Speeding Infrastructure Development through More Efficient and Effective Permitting and Environmental Review), Executive Order 13604 of March 22, 2012 (Improving Performance of Federal Permitting and Review of Infrastructure Projects), and my memorandum of May 17, 2013 (Modernizing Federal Infrastructure Review and Permitting Regulations, Policies, and Procedures); and

(iv) report on their progress in achieving the requirements identified above, including accomplished and planned milestones, in the Agency Adaptation Plans developed pursuant to section 5 of this order.

. . .

Sec. 3. Managing Lands and Waters for Climate Preparedness and Resilience. Within 9 months of the date of this order and in coordination with the efforts described in section 2 of this order, the heads of the Departments of Defense, the Interior, and Agriculture, the Environmental Protection Agency, NOAA, the Federal Emergency Management Agency, the Army Corps of Engineers, and other agencies as recommended by the Council established in section 6 of this order shall work with the Chair of CEQ and the Director of the Office of Management and Budget (OMB) to complete an inventory and assessment of proposed and completed changes to their land- and water-related policies, programs, and regulations necessary to make the Nation's watersheds, natural resources, and ecosystems, and the communities and economies that depend on them, more resilient in the face of a changing climate.

. . .

Sec. 4. Providing Information, Data, and Tools for Climate Change Preparedness and Resilience. (a) In support of Federal, regional, State, local, tribal, private-sector and nonprofit-sector efforts to prepare for the impacts of climate change, the Departments of Defense, the Interior, Agriculture, Commerce, Health and Human Services, Housing and Urban Development, Transportation, Energy, and Homeland Security, the Environmental Protection Agency, the National Aeronautics and Space Administration, and any other agencies as recommended by the Council established in section 6 of this order, shall, supported by USGCRP, work together to develop and provide authoritative, easily accessible, usable, and timely data, information, and decision-support tools on climate preparedness and resilience.

. . .

Sec. 5. Federal Agency Planning for Climate Change Related Risk. (a) Consistent with Executive Order 13514, agencies have developed Agency Adaptation Plans and provided them to CEQ and OMB. These plans evaluate the most significant climate change related risks to, and vulnerabilities in, agency operations and missions in both the short and long term, and outline actions that agencies will take to manage these risks and vulnerabilities. Building on these efforts, each agency shall develop or continue to develop, implement, and update comprehensive plans that integrate consideration of climate change into agency operations and overall mission objectives and submit those plans to CEQ and OMB for review. Each Agency Adaptation Plan shall include:

[The next sections list the items to be included in each plan.]

. . .

Sec. 6. Council on Climate Preparedness and Resilience.

(a) Establishment. There is established an interagency Council on Climate Preparedness and Resilience (Council).

[The members of the Council are listed next.]

. . .

Sec. 7. State, Local, and Tribal Leaders Task Force on Climate Preparedness and Resilience.

(a) Establishment. To inform Federal efforts to support climate preparedness and resilience, there is established a State, Local, and Tribal Leaders Task Force on Climate Preparedness and Resilience (Task Force).

[Details of the membership and responsibilities of the Task Force are listed next.]

. . .

BARACK OBAMA

Source: "Executive Order—Preparing the United States for the Impacts of Climate Change." 2013. The White House. https:// obamawhitehouse.archives.gov/the-press-office/2013/11/01

/executive-order-prEPAring-united-states-impacts-climate
-change. Accessed on August 6, 2019.

The Clean Power Plan (2015)

*The most common approach for dealing with climate change ad-
opted by the federal, state, local, and other governmental units in
the United States is reduction in dependence on fossil fuels for energy generation. In 2015, the administration of President Barack
Obama issued the Clean Power Plan, an effort to achieve this goal
by cutting back on support for coal, oil, and other fossil fuel use in
the United States. The selection below provides a summary of the
main features of that plan.*

Clean Power Plan

CUTTING CARBON POLLUTION FROM POWER PLANTS

. . . *[Introduction]*

What is the Clean Power Plan?

- The Clean Power Plan will reduce carbon pollution from
  power plants, the nation's largest source, while maintaining
  energy reliability and affordability. Also on August 3, EPA
  issued final Carbon Pollution Standards for new, modified,
  and reconstructed power plants, and proposed a Federal
  Plan and model rule to assist states in implementing the
  Clean Power Plan.
- These are the first-ever national standards that address carbon pollution from power plants.
- The Clean Power Plan cuts significant amounts of power
  plant carbon pollution and the pollutants that cause the soot
  and smog that harm health, while advancing clean energy
  innovation, development and deployment, and laying the

foundation for the long-term strategy needed to tackle the threat of climate change. By providing states and utilities ample flexibility and the time needed to achieve these pollution cuts, the Clean Power Plan offers the power sector the ability to optimize pollution reductions while maintaining a reliable and affordable supply of electricity for ratepayers and businesses.

- Fossil fuels will continue to be a critical component of America's energy future. The Clean Power Plan simply makes sure that fossil fuel-fired power plants will operate more cleanly and efficiently, while expanding the capacity for zero- and low-emitting power sources.

- The final rule is the result of unprecedented outreach to states, tribes, utilities, stakeholders and the public, including more than 4.3 million comments EPA received on the proposed rule. The final Clean Power Plan reflects that input, and gives states and utilities time to preserve ample, reliable and affordable power for all Americans.

Why We Need the Clean Power Plan

- In 2009, EPA determined that greenhouse gas pollution threatens Americans' health and welfare by leading to long-lasting changes in our climate that can have a range of negative effects on human health and the environment. Carbon dioxide (CO_2) is the most prevalent greenhouse gas pollutant, accounting for nearly three-quarters of global greenhouse gas emissions and 82 percent of U.S. greenhouse gas emissions.

- Climate change is one of the greatest environmental and public health challenges we face. Climate impacts affect all Americans' lives—from stronger storms to longer droughts and increased insurance premiums, food prices, and allergy seasons.

- 2014 was the hottest year in recorded history, and 14 of the 15 warmest years on record have all occurred in the first

15 years of this century. Recorded temperatures in the first half of 2015 were also warmer than normal.

- Overwhelmingly, the best scientists in the world, relying on troves of data and millions of measurements collected over the course of decades on land, in air and water, at sea and from space, are telling us that our activities are causing climate change.

- The most vulnerable among us—including children, older adults, people with heart or lung disease and people living in poverty—may be most at risk from the impacts of climate change.

- Fossil fuel-fired power plants are by far the largest source of U.S. CO_2 emissions, making up 31 percent of U.S. total greenhouse gas emissions.

- Taking action now is critical. Reducing CO_2 emissions from power plants, and driving investment in clean energy technologies strategies that do so, is an essential step in lessening the impacts of climate change and providing a more certain future for our health, our environment, and future generations.

Benefits of Implementing the Clean Power Plan

- The transition to clean energy is happening faster than anticipated. This means carbon and air pollution are already decreasing, improving public health each and every year.

- The Clean Power Plan accelerates this momentum, putting us on pace to cut this dangerous pollution to historically low levels in the future.

- When the Clean Power Plan is fully in place in 2030, carbon pollution from the power sector will be 32 percent below 2005 levels, securing progress and making sure it continues.

- The transition to cleaner sources of energy will better protect Americans from other harmful air pollution, too. By 2030, emissions of sulfur dioxide from power plants will be

90 percent lower compared to 2005 levels, and emissions of nitrogen oxides will be 72 percent lower. Because these pollutants can create dangerous soot and smog, the historically low levels mean we will avoid thousands of premature deaths and have thousands fewer asthma attacks and hospitalizations in 2030 and every year beyond.

- Within this larger context, the Clean Power Plan itself is projected to contribute significant pollution reductions, resulting in important benefits, including:
  ○ Climate benefits of $20 billion
  ○ Health benefits of $14–$34 billion
  ○ Net benefits of $26–$45 billion
- Because carbon pollution comes packaged with other dangerous air pollutants, the Clean Power Plan will also protect public health, avoiding each year:
  ○ 3,600 premature deaths
  ○ 1,700 heart attacks
  ○ 90,000 asthma attacks
  ○ 300,000 missed work days and school days

How the Clean Power Plan Works

- The Clean Air Act—under section 111(d)—creates a partnership between EPA, states, tribes and U.S. territories—with EPA setting a goal and states and tribes choosing how they will meet it.
- The final Clean Power Plan follows that approach. EPA is establishing interim and final carbon dioxide (CO_2) emission performance rates for two subcategories of fossil fuel-fired electric generating units (EGUs):
  ○ Fossil fuel-fired electric steam generating units (generally, coal- and oil-fired power plants)
  ○ Natural gas-fired combined cycle generating units

- To maximize the range of choices available to states in implementing the standards and to utilities in meeting them, EPA is establishing interim and final statewide goals in three forms:

  ○ A rate-based state goal measured in pounds per megawatt hour (lb/MWh);

  ○ A mass-based state goal measured in total short tons of CO_2;

  ○ A mass-based state goal with a new source complement measured in total short tons of CO_2.

- States then develop and implement plans that ensure that the power plants in their state—either individually, together or in combination with other measures—achieve the interim CO_2 emissions performance rates over the period of 2022 to 2029 and the final CO_2 emission performance rates, rate-based goals or mass-based goals by 2030.

- These final guidelines are consistent with the law and align with the approach that Congress and EPA have always taken to regulate emissions from this and all other industrial sectors—setting source-level, source category-wide standards that sources can meet through a variety of technologies and measures.

Source: "FACT SHEET: Overview of the Clean Power Plan." 2015. EPA. https://archive.epa.gov/epa/cleanpowerplan/fact -sheet-overview-clean-power-plan.html. Accessed on August 15, 2019.

Executive Order 13783 (2017)

The federal government has relatively few laws, regulations, or other official documents dealing with climate change. Those documents that do exist generally reflect the position of one presidential administration or another about the topic and/or congressional actions on the subject. As an example, President Donald Trump had a substantially different view of climate change than had his predecessor, Barack Obama, a position he made clear before his election in 2016. He had, in fact, expressed the view that climate change was a political hoax perpetrated on the world by the Chinese government. At one point in his political campaign, for example, he also noted: "Among the lowest temperatures EVER in much of the United States. Ice caps at record size. Changed name from GLOBAL WARMING to CLIMATE CHANGE" (Matthews, Dylan. 2017. "Donald Trump Has Tweeted Climate Change Skepticism 115 Times. Here's All of It." Vox. https://www .vox.com/policy-and-politics/2017/6/1/15726472/trump-tweets -global-warming-paris-climate-agreement. Accessed on August 6, 2019). Early in his presidency, Trump reversed several policy actions taken by his predecessors, primarily Obama. An excerpt from that Executive Order follows.

[The order begins with a statement of the president's philosophy about energy production in the United States.]

Section 1. Policy. (a) It is in the national interest to promote clean and safe development of our Nation's vast energy resources, while at the same time avoiding regulatory burdens that unnecessarily encumber energy production, constrain economic growth, and prevent job creation. Moreover, the prudent development of these natural resources is essential to ensuring the Nation's geopolitical security.

(b) It is further in the national interest to ensure that the Nation's electricity is affordable, reliable, safe, secure, and clean, and that it can be produced from coal, natural gas, nuclear material, flowing water, and other domestic sources, including renewable sources.

(c) Accordingly, it is the policy of the United States that executive departments and agencies (agencies) immediately review existing regulations that potentially burden the development or use of domestically produced energy resources and appropriately suspend, revise, or rescind those that unduly burden the development of domestic energy resources beyond the degree necessary to protect the public interest or otherwise comply with the law.

. . .

[He then instructs members of the administration to review existing policy in light of the previous statement of philosophy.]

Sec. 2. Immediate Review of All Agency Actions that Potentially Burden the Safe, Efficient Development of Domestic Energy Resources. (a) The heads of agencies shall review all existing regulations, orders, guidance documents, policies, and any other similar agency actions (collectively, agency actions) that potentially burden the development or use of domestically produced energy resources, with particular attention to oil, natural gas, coal, and nuclear energy resources. Such review shall not include agency actions that are mandated by law, necessary for the public interest, and consistent with the policy set forth in section 1 of this order.

(b) For purposes of this order, "burden" means to unnecessarily obstruct, delay, curtail, or otherwise impose significant costs on the siting, permitting, production, utilization, transmission, or delivery of energy resources.

. . .

[Next, the president lists the specific actions of preceding administrations to be revoked by way of this Executive Order.]

Sec. 3. Rescission of Certain Energy and Climate-Related Presidential and Regulatory Actions. (a) The following Presidential actions are hereby revoked:

(i) Executive Order 13653 of November 1, 2013 (Preparing the United States for the Impacts of Climate Change);

(ii) The Presidential Memorandum of June 25, 2013 (Power Sector Carbon Pollution Standards);

(iii) The Presidential Memorandum of November 3, 2015 (Mitigating Impacts on Natural Resources from Development and Encouraging Related Private Investment); and

(iv) The Presidential Memorandum of September 21, 2016 (Climate Change and National Security).

(b) The following reports shall be rescinded:

(i) The Report of the Executive Office of the President of June 2013 (The President's Climate Action Plan); and

(ii) The Report of the Executive Office of the President of March 2014 (Climate Action Plan Strategy to Reduce Methane Emissions).

(c) The Council on Environmental Quality shall rescind its final guidance entitled "Final Guidance for Federal Departments and Agencies on Consideration of Greenhouse Gas Emissions and the Effects of Climate Change in National Environmental Policy Act Reviews," which is referred to in "Notice of Availability," 81 Fed. Reg. 51866 (August 5, 2016).

(d) The heads of all agencies shall identify existing agency actions related to or arising from the Presidential actions listed in subsection (a) of this section, the reports listed in subsection (b) of this section, or the final guidance listed in subsection (c) of this section. Each agency shall, as soon as practicable, suspend, revise, or rescind, or publish for notice and comment proposed rules suspending, revising, or rescinding any such actions, as appropriate and consistent with law and with the policies set forth in section 1 of this order.

Sec. 4. Review of the Environmental Protection Agency's "Clean Power Plan" and Related Rules and Agency Actions. (a) The Administrator of the Environmental Protection Agency

(Administrator) shall immediately take all steps necessary to review the final rules set forth in subsections (b)(i) and (b)(ii) of this section, and any rules and guidance issued pursuant to them, for consistency with the policy set forth in section 1 of this order and, if appropriate, shall, as soon as practicable, suspend, revise, or rescind the guidance, or publish for notice and comment proposed rules suspending, revising, or rescinding those rules. In addition, the Administrator shall immediately take all steps necessary to review the proposed rule set forth in subsection (b)(iii) of this section, and, if appropriate, shall, as soon as practicable, determine whether to revise or withdraw the proposed rule.

(b) This section applies to the following final or proposed rules:

(i) The final rule entitled "Carbon Pollution Emission Guidelines for Existing Stationary Sources: Electric Utility Generating Units," 80 Fed. Reg. 64661 (October 23, 2015) (Clean Power Plan);

(ii) The final rule entitled "Standards of Performance for Greenhouse Gas Emissions from New, Modified, and Reconstructed Stationary Sources: Electric Utility Generating Units," 80 Fed. Reg. 64509 (October 23, 2015); and

(iii) The proposed rule entitled "Federal Plan Requirements for Greenhouse Gas Emissions from Electric Utility Generating Units Constructed on or Before January 8, 2014; Model Trading Rules; Amendments to Framework Regulations; Proposed Rule," 80 Fed. Reg. 64966 (October 23, 2015).

(c) The Administrator shall review and, if appropriate, as soon as practicable, take lawful action to suspend, revise, or rescind, as appropriate and consistent with law, the "Legal Memorandum Accompanying Clean Power Plan for Certain Issues," which was published in conjunction with the Clean Power Plan.

(d) The Administrator shall promptly notify the Attorney General of any actions taken by the Administrator pursuant to this order related to the rules identified in subsection (b) of this section so that the Attorney General may, as appropriate, provide notice of this order and any such action to any court with jurisdiction over pending litigation related to those rules, and may, in his discretion, request that the court stay the litigation or otherwise delay further litigation, or seek other appropriate relief consistent with this order, pending the completion of the administrative actions described in subsection (a) of this section.

Sec. 5. Review of Estimates of the Social Cost of Carbon, Nitrous Oxide, and Methane for Regulatory Impact Analysis. (a) In order to ensure sound regulatory decision making, it is essential that agencies use estimates of costs and benefits in their regulatory analyses that are based on the best available science and economics.

(b) The Interagency Working Group on Social Cost of Greenhouse Gases (IWG), which was convened by the Council of Economic Advisers and the OMB Director, shall be disbanded, and the following documents issued by the IWG shall be withdrawn as no longer representative of governmental policy:

(i) Technical Support Document: Social Cost of Carbon for Regulatory Impact Analysis under Executive Order 12866 (February 2010);

(ii) Technical Update of the Social Cost of Carbon for Regulatory Impact Analysis (May 2013);

(iii) Technical Update of the Social Cost of Carbon for Regulatory Impact Analysis (November 2013);

(iv) Technical Update of the Social Cost of Carbon for Regulatory Impact Analysis (July 2015);

(v) Addendum to the Technical Support Document for Social Cost of Carbon: Application of the Methodology to Estimate the Social Cost of Methane and the Social Cost of Nitrous Oxide (August 2016); and

(vi) Technical Update of the Social Cost of Carbon for Regulatory Impact Analysis (August 2016).

[Finally, the President makes clear his administration's position on the generation of energy using fossil fuels in contrast to renewable fuels.]

Sec. 6. Federal Land Coal Leasing Moratorium. The Secretary of the Interior shall take all steps necessary and appropriate to amend or withdraw Secretary's Order 3338 dated January 15, 2016 (Discretionary Programmatic Environmental Impact Statement (PEIS) to Modernize the Federal Coal Program), and to lift any and all moratoria on Federal land coal leasing activities related to Order 3338. The Secretary shall commence Federal coal leasing activities consistent with all applicable laws and regulations.

Sec. 7. Review of Regulations Related to United States Oil and Gas Development. (a) The Administrator shall review the final rule entitled "Oil and Natural Gas Sector: Emission Standards for New, Reconstructed, and Modified Sources," 81 Fed. Reg. 35824 (June 3, 2016), and any rules and guidance issued pursuant to it, for consistency with the policy set forth in section 1 of this order and, if appropriate, shall, as soon as practicable, suspend, revise, or rescind the guidance, or publish for notice and comment proposed rules suspending, revising, or rescinding those rules.

(b) The Secretary of the Interior shall review the following final rules, and any rules and guidance issued pursuant to them, for consistency with the policy set forth in section 1 of this order and, if appropriate, shall, as soon as practicable, suspend, revise, or rescind the guidance, or publish for notice and comment proposed rules suspending, revising, or rescinding those rules:

(i) The final rule entitled "Oil and Gas; Hydraulic Fracturing on Federal and Indian Lands," 80 Fed. Reg. 16128 (March 26, 2015);

(ii) The final rule entitled "General Provisions and Non-Federal Oil and Gas Rights," 81 Fed. Reg. 77972 (November 4, 2016);

(iii) The final rule entitled "Management of Non Federal Oil and Gas Rights," 81 Fed. Reg. 79948 (November 14, 2016); and

(iv) The final rule entitled "Waste Prevention, Production Subject to Royalties, and Resource Conservation," 81 Fed. Reg. 83008 (November 18, 2016).

. . .

DONALD J. TRUMP

THE WHITE HOUSE,

March 28, 2017.

Source: "Presidential Executive Order on Promoting Energy Independence and Economic Growth." March 28, 2017. https://www.whitehouse.gov/presidential-actions/presidential -executive-order-promoting-energy-independence-economic -growth/. Accessed on August 6, 2019.

The Affordable Clean Energy Rule (2019)

Four years after Obama's Clean Power Plan, the administration of President Donald Trump set forth a somewhat different plan for dealing with energy generation. It rescinded the Clean Power Plan and substituted its own ideas in the Affordable Clean Energy (ACE) Rule. This rule argued that alternative sources of energy, such as wind and solar, were not adequate to meet the nation's energy needs and that the federal government should continue its support for fossil fuel use, albeit with efforts to improve the efficiency of those fuels. The selection below provides a summary of the Trump administration plan.

. . . [Introduction]

PROMULGATING ACE AND ESTABLISHING EMISSION GUIDELINES

- ACE establishes heat rate improvement (HRI), or efficiency improvement, as the best system of emissions reduction (BSER) for CO_2 from coal-fired EGUs.
- Heat rate is a measure of the amount of energy required to generate a unit of electricity.
- By employing a broad range of HRI technologies and techniques, EGUs can more efficiently generate electricity with less carbon intensity.
- An improvement to heat rate results in a reduction in the emission rate of an EGU (in terms of CO_2 emissions per unit of electricity produced).
- The BSER is the best technology or other measure that has been adequately demonstrated to improve emissions performance for a specific industry or process (a "source category"). In determining the BSER, EPA considers technical feasibility, cost, non-air quality health and environmental impacts, and energy requirements.
- The BSER must be applicable to, at, and on the premises of an affected facility.
- ACE lists six HRI "candidate technologies," as well as additional operating and maintenance practices.
- The six candidate technologies are:
  - Neural Network/Intelligent Sootblowers
  - Boiler Feed Pumps
  - Air Heater and Duct Leakage Control

- Variable Frequency Drives
- Blade Path Upgrade (Steam Turbine)
- Redesign/Replace Economizer
- For each candidate technology, EPA has provided information regarding the degree of emission limitation achievable through application of the BSER as ranges of expected improvement and costs.
- States will establish unit-specific "standards of performance" that reflect the emission limitation achievable through application of the BSER technologies.
- These technologies, equipment upgrades, and best operating and maintenance practices were determined to comprise the BSER because they can be applied broadly and are expected to provide significant HRI without limitations due to geography, fuel type, etc.
- ACE recognizes that EPA's statutory role is to determine the BSER and the degree of emission limitation achievable through application of the BSER, and that the states' role is to develop plans that establish unit-specific standards of performance that reflect application of the BSER.
- The CAA provides that states establish the standards of performance and explicitly directs EPA to allow states to consider "the remaining useful life of the source" and other source-specific factors in establishing standards of performance.
- States will evaluate applicability to their existing sources of the six candidate technologies and improved operating and maintenance practices and take into consideration source-specific factors in establishing a standard of performance at the unit level.
- States will submit plans to EPA that establish standards of performance and include measures that provide for the implementation and enforcement of such standards.

- The plan submissions must explain how the state applied the BSER to each source—and how the state took other factors into consideration—in setting unit-specific standards. These plans are due in three years.

Source: "FACT SHEET: The Affordable Clean Energy Rule (ACE)." 2019. EPA. https://www.epa.gov/stationary-sources -air-pollution/fact-sheet-overview-final-ace-rule. Accessed on August 15, 2019.

Climate change has been the subject of many books, articles, reports, and electronic sources for more than five decades. No single bibliography can adequately cover all, or even most, of the items. This bibliography is designed to give the reader a taste of the resources available on the topic. In some cases, a resource may be available in two different formats, printed article and online version of the article, for example. In such cases the availability of the resource in both media is indicated in the citation. In addition to the items listed here, the reader is encouraged to review the resources listed at the end of Chapters 1 and 2 to find suggestions for additional readings.

Books

Acred, Cara. 2015. *Climate Change Debate.* Cambridge, UK: Independence.

> This book is volume 286 in the publisher's "Issues" series on important topics in the world today. It consists of facts, articles, and opinion pieces from print and

---

Flooding in a village on the island of Charkajal, Gulf of Bengal, Bangladesh. As a result of recent floodings, villagers sometimes have to sit for days on a mound or islet, waiting with their cattle for the water to drop. Global warming has caused the country to become more and more flooded by river waters, the ascent of the sea level, and the heavy rainfall during the monsoon. It is often the poorest people who are most affected by these natural disasters. (Sjors737/Dreamstime.com)

electronic media, along with reports from governmental and non-governmental organizations and special interest groups.

Banholzer, Sandra, James Kossin, and Simon Donner. 2014. "The Impact of Climate Change on Natural Disasters." In *Reducing Disaster: Early Warning Systems for Climate Change*, edited by Ashbindu Singh and Zinta Zommers, ch. 2, 21–49. Dordrecht: Springer. https://www.ssec.wisc.edu/~kossin/articles/Chapter_2.pdf. Accessed on September 23, 2019.

This chapter discusses past instances of the impact of human activities on natural disasters and likely prospects for such events in the future. It provides a very good introduction to the field of attribution science.

Blackstock, Jason J., and Sean Low, eds. 2019. *Geoengineering Our Climate? Ethics, Politics and Governance*. Abingdon, UK and New York: Routledge.

The papers in this anthology discuss all aspects of geoengineering of the climate. They include articles on the history of geoengineering proposals, contemporary issues and attitudes about geoengineering, early experiments in the field, existing institutions for dealing with geoengineering issues, national and regional perspectives, and governance of the technology.

Bodansky, Daniel. 2001. "The History of the Global Climate Change Regime." In *International Relations and Global Climate Change*, edited by Urs Luterbacher and Detlef F. Sprinz, 23–40. Cambridge, MA: MIT Press. http://www.oyetimes.com/wp-content/uploads/2018/01/luterbacher20chapter20220102.pdf. Accessed on September 23, 2019.

The 1980s were an important period during which researchers began to promote concern about climate change. This chapter provides an excellent review of the most important events that occurred during this period, along with their impact on climate change debates of the future.

Boulter, Sarah, et al. 2018. *Natural Disasters and Adaptation to Climate Change*. Cambridge, UK: Cambridge University Press.
The authors explore specific natural disasters for which some connection with climate change have been claimed. The four sections include examples from North America, Australia, Europe, and the developing world. A final section provides a synthesis gained from studying these specific examples.

Bradley, Raymond S. 2015. *Paleoclimatology: Reconstructing Climates of the Quaternary*. Amsterdam: Elsevier/Academic Press.
The author explains and discusses essentially all of the proxies available for studying climatic trends during the Quaternary Period.

Dawson, Alastair G. 2016. *Ice Age Earth: Late Quaternary Geology and Climate*. New York: Routledge.
This book provides a good, comprehensive, if somewhat technical review of almost every aspect of global climate changes during the Quaternary period, with sections on ocean sediments and ice cores; the melting of the last great ice sheets; lakes, bogs, and mires; Late Quaternary volcanic activity, and crustal and subcrustal events.

Dressler, Andrew, and Edward A. Parson. 2019. *The Science and Politics of Global Climate Change: A Guide to the Debate*. 2nd ed. Cambridge, UK and New York: Cambridge University Press.
The authors describe their text as a work that "provides an integrated treatment of the science, technology, economics, policy, and politics of climate change for the educated non-specialist reader."

Fløttum, Kjersti, ed. 2019. *The Role of Language in the Climate Change Debate*. New York and London: Routledge.
A critical element in discussions about climate change is the way in which opinions are expressed. Words make a difference! The essays in this book provide a somewhat technical analysis of the way language is used by individuals on all sides of the climate change debate and the effects this has on an understanding of and action (or inaction) on the problem.

Goldstein, Joshua S., and Staffan A. Qvist. 2019. *A Bright Future: How Some Countries Have Solved Climate Change and the Rest Can Follow*. New York: Public Affairs.
Many commentaries on climate change focus on the possible disastrous effects that may result from the phenomenon. The authors of this book point out that some parts of the world have already converted to low-carbon technologies, with little or no harm, and often substantial benefit, to their own economies. They say that other nations, states, and regions can do the same.

Gornitz, Vivien, ed. [2009] 2015. *Encyclopedia of Paleoclimatology and Ancient Environments*. Dordrecht, Netherlands, and New York: Springer; Boston, MA: Credo Reference.
This collection of more than 200 entries covers a vast range of topics related to climate change and its effects on the natural and human environments.

Griggs, Gary B. 2017. *Coasts in Crisis: A Global Challenge*. Oakland: University of California Press.
This book explores the special risks posed for coastal areas around the world from natural storms, human activities, and climate change.

Grove, Jean M. 2015. *The Little Ice Age*. London: Routledge.
This book focuses on the characteristics of the Little Ice Age in several parts of the planet during its presence.

Hallegatte, Stéphane. 2016. *Natural Disasters and Climate Change*. Cham, Switzerland: Springer.

The author explores the evidence for a connection between natural disasters and climate change and some consequences of this new relationship in today's world. Some of the topics considered are the meaning of a disaster from an economic point of view, what is the evidence for the concept of a "disaster risk," what are the general trends in the effects of climate change on natural disasters, methodologies for managing natural disasters in light of climate change, and elements of decision making about natural disasters in a changing world.

Hamin, Elisabeth M., Yaser Abunnasr, and Robert L. Ryan. 2019. *Planning for Climate Change: A Reader in Green Infrastructure and Sustainable Design for Resilient Cities*. New York: Routledge.

Climate change is almost certain to affect various natural and human environments in different ways. This book focuses on the unique effects of climate change on urban areas, the types of effects to be expected, and mechanisms that can be developed for dealing with these effects.

Hannah, Lee Jay. 2011. *Climate Change Biology*. Burlington, MA: Academic Press.

This book is devoted to the emerging discipline of biological changes that occur as a result of climate change, especially anthropogenic climate change. It includes topics such as changes in species ranges; changes in timing of biological events; changes in ecosystems; past changes in marine, freshwater, and land ecosystems; extinctions; species management; and mitigation of effects.

Hay, William W. 2016. *Experimenting on a Small Planet: A History of Scientific Discoveries, a Future of Climate Change and Global Warming*. Cham, Switzerland: Springer.

The author discusses most of the basic topics in the field of climate science, such as the language of science, geologic time, documenting past climate change, albedo, oxygen and ozone, carbon dioxide, sea level, and global and regional aspects of climate change.

Hoffman, Andrew J. 2015. *How Culture Shapes the Climate Change Debate*. Stanford, CA: Stanford Briefs.
The author attempts to identify cultural factors that help explain the often wide gap between the understanding, acceptance, and appreciation of climate change by different groups and individuals. Book chapters include "A Cultural Schism," "Social Psychology and the Climate Change Debate," "Sources of Organized Resistance," and "Historical Analogies for Cultural Change."

Humphreys, Stephen, ed. 2009. *Human Rights and Climate Change*. Leiden: Cambridge University Press.
Climate change has the potential for affecting human rights in many areas, including health care, corporate accountability, inequities in risk, and forest resources. The essays in this book expand on each of these and other effects of climate change on human rights.

Inhofe, James. 2012. *The Greatest Hoax: How the Global Warming Conspiracy Threatens Your Future*. New York: WND Books.
Probably one of the best-known books arguing that global warming and climate changes are "hoaxes," this book is a must reading for an understanding of the basic arguments of climate change skeptics and deniers.

Islam, R. I., and M. M. Khan. 2019. *The Science of Climate Change*. Beverly, MA: Scrivener; Hoboken, NJ: John Wiley.
This book provides a broad, general, heavily illustrated introduction to the issue of climate change, with chapters on forest fires and anthropogenic carbon dioxide, agricultural contributions to climate change, biofuel production

and refining processes and global warming, the history of climate change hysteria, and the monetization of climate science.

Kelbaugh, Doug. 2019. *The Urban Fix: Resilient Cities in the War against Climate Change, Heat Islands and Overpopulation.* New York: Routledge.

The author reviews the special characteristics of urban areas that relate to climate change, the challenges these features have for residents of the areas, and possible solutions for dealing with cities of the future in an age of global warming.

Leonard, Christopher. 2019. *Kochland: The Secret History of Koch Industries and Corporate Power in America.* London: Simon & Schuster.

A great deal has been written about the influence of individual business executives on efforts to deal with climate change, usually with the goal of discrediting the findings of research on the subject. The Koch brothers, Charles and David, have historically been among the most active individuals in this field. This book describes their efforts over the years to discredit climate change research and its possible effects on the environment and economy.

Letcher, Trevor M., ed. 2019. *Managing Global Warming: An Interface of Technology and Human Issues.* London: United Kingdom Academic Press.

This book provides a review of alternative energy sources that can be used to reduce the world's dependence on fossil fuels.

McKibben, Bill. 2012. *The Global Warming Reader: A Century of Writing about Climate Change.* New York: Penguin Books.

This collection of nearly four dozen historic articles about climate change includes selections by Arrhenius, Callendar, Keeling, Hansen, Gore, Inhofe, and Crichton.

Nagel, Joane. 2016. *Gender and Climate Change: Impacts, Science, Policy.* New York: Routledge.

Nagel notes that some natural disasters affect all individuals, regardless of their genders. She observes that an increasing body of research indicates that such events *do* have different effects on different genders. She explores the relationships of gender and global warming, sea-level rise, climate change science, the military-science complex, climate change skepticism, and climate change policy.

Palmer, Lisa. 2017. *Hot, Hungry Planet: The Fight to Stop a Global Food Crisis in the Face of Climate Change.* New York: St. Martin's Press.

Food security is a major issue of concern in discussions about climate change. In this book, Palmer examines the changes that are likely to occur in agriculture as a result of climate changes, the challenges these changes will bring for feeding the world's population, and some solutions for these problems.

Philander, S. George. 2012. *Encyclopedia of Global Warming & Climate Change.* 3 vols. Thousand Oaks, CA: SAGE.

This work includes entries on virtually every aspect of the topics of global warming and climate change that one might imagine. An invaluable resource for the researchers.

Preston, Christopher J. 2016. *Climate Justice and Geoengineering: Ethics and Policy in the Atmospheric Anthropocene.* London: Rowan and Littlefield, International.

Much of the discussion about geoengineering for climate change is based on technical issues: How would it be used? How successful is it likely to be? What are the technical problems in making it work? This book focuses instead on a variety of ethical issues, such as whether researchers are ethically justified or required to use geoengineering to "save the world," how much they are obligated to develop the technology as a form of compensation to the poor, how geoengineering would be related to Earth's food security problems, and what the issues are for the fair distribution of geoengineering technologies.

Rapp, Donald. 2019. *Ice Ages and Interglacials: Measurements, Interpretation, and Models.* Cham, Switzerland: Springer.
This excellent text provides a broad, detailed, and technical introduction to the topic of ice ages and interglacials. Topics included are the history and description of the ice ages, factors that affect Earth's climate, methodology of ice coring, ice age proxies, models of the ice ages, and how the ice ages ended.

Rodhe, Henning, and Robert Charlson, eds. 1998. *Svante Arrhenius and the Greenhouse Effect.* Stockholm: Royal Swedish Academy.
This collection of essays was prepared to commemorate the 100th anniversary of Arrhenius's article, "On the Influence of Carbonic Acid in the Air upon the Temperature of the Ground." Articles cover topics such as Arrhenius and the Greenhouse Effect, Carbon Dioxide and Warming of the Early Earth, Palaeoclimate Sensitivity to CO_2 and Insolation, and From Arrhenius to Megascience: Interplay between Science and Public Decisionmaking.

Skillington, Tracey. 2019. *Climate Change and Intergenerational Justice.* London: Routledge.
Climate change issues affect individuals very differently depending on their generation. For young people, the topic is one of intense concern for their future, while members of the older generation may tend to regard such issues as problems that can be handled in the future. This book reviews the way in which members of different generations view climate change topics and ways in which they can be assessed and handled.

Thomas, Vinod. 2018. *Climate Change and Natural Disasters: Transforming Economies and Policies for a Sustainable Future.* New York: Routledge.

This book deals with various aspect of the relationship between climate change and natural disasters. Some of the topics considered include the anthropogenic link between climate change and natural disasters, the rising threat of such events to various parts of the world, methods of climate change mitigation, and adaptation to a changing climate and management of anthropogenic natural disaster events.

Articles

Some journals are devoted entirely or in large measure to articles about climate change. Some of the journals are as follows:

Carbon Balance and Management: ISSN: 1750-0680 (online)

Climate and Development: ISSN: 1756-5529 (print); 1756-5537 (online)

Climatic Change: ISSN: 0165-0009 (print); 1573-1480 (online)

Climate Dynamics: ISSN: 0930-7575 (print); 1432-0894 (online)

Climate of the Past: ISSN: 1814-9324 (print); 1814-9332 (online)

Climate Policy: ISSN: 1469-3062 (print); 1752-7457 (online)

Climate Research: ISSN: 0936-577X (print); 1616-1572 (online)

Global and Planetary Change: ISSN: 0921-8181 (online)

Global Change Biology: ISSN: 1365-2486 (online)

International Journal of Climatology: ISSN: 1097-0088 (online)

Nature Climate Change: ISSN: 1758-6798 (online)

PAGES Magazine: ISSN: 2411-605X print); 2411-9180 (online)

Weather, Climate, and Society: ISSN: 1948-8327 (print); 1948-8335 (online)

Anderson, Christa M., et al. 2019. "Natural Climate Solutions Are Not Enough." *Science* 363(6430): 933–934. doi:10.1126 /science.aaw2741.

The authors review and comment on several climate change solutions that involve manipulating the natural environment in one way or another. They then argue that this limited approach to the problem is not sufficient for solving future problems and that some form of carbon control is necessary to deal with the issue.

Anderson, David G., et al. 2017. "Sea-Level Rise and Archaeological Site Destruction: An Example from the Southeastern United States Using DINAA (Digital Index of North American Archaeology)." *PLoS One* 12(11): e0188142. https://journals .plos.org/plosone/article?id=10.1371/journal.pone.0188142. Accessed on September 27, 2019.

One of the many fields of science in which climate change may have important long-term effects is archaeology. The science depends, of course, on relics buried underground or in shallow waters that can be recovered and studied by researchers. Recently, increasing concerns have been expressed about the loss of important artifacts as a consequence of global warming. This article reports on just one of these warnings. Also see Holleson, Matthiesen, and Elberling 2017.

Barnes, David K. A., and Geraint A. Tarling. 2017. "Polar Oceans in a Changing Climate." *Current Biology* 27(11): R454–R460.https://www.cell.com/action/showPdf?pii=S0960 -9822%2817%2930080-5. Accessed on September 23, 2019.

This article provides a very understandable introduction to the nature of life in the polar oceans and changes that have and are taking place in those settings as a result of climate change.

Barnola, J.-M., et al. 2003. "Historical CO_2 Record from the Vostok Ice Core." In "Trends: A Compendium of Data on Global Change." Carbon Dioxide Information Analysis Center, Oak Ridge National Laboratory, U.S. Department of Energy. https://cdiac.ess-dive.lbl.gov/trends/co2/vostok.html. Accessed on September 20, 2019.

> Ice cores collected at the Vostok Research Center in Antarctica provide some of the most revealing data available about global temperatures over the past ~400,000 years. This article describes how the research has been done and provides data that have been collected in the studies.

Bedford, Daniel. 2010. "Agnotology as a Teaching Tool: Learning Climate Science by Studying Misinformation." *Journal of Geography* 109(4): 159–165.

> The author argues that the dissemination of misinformation about climate change is a major factor for the general public's "confusion" about the topic. He describes a course in which he uses this misinformation as a way for teaching correct information about the topic. The paper is of considerable interest because of the reactions it produced in the literature. For a review of that controversy, see John Cook. 2013. "New Paper on Agnotology and Scientific Consensus." Skeptical Science. https://skepticalscience.com/New-paper-agnotology-scientific-consensus.html. Accessed on September 23, 2019. (Agnotology is defined as the study of ignorance and how it is produced.)

Charlson, Robert J. 1997. "Direct Climate Forcing by Anthropogenic Sulfate Aerosols: The Arrhenius Paradigm a Century Later." *Ambio* 26(1): 25–31.

> The author refers to the early research of Arrhenius, Ångström, and Bergeron to review the ways in which aerosol particles, such as those produced by volcanic eruptions, influence climate.

Easterling, David R., et al. 2016. "Detection and Attribution of Climate Extremes in the Observed Record." *Weather and*

Climate Extremes 11: 17–27. doi:10.1016/j.wace.2016.01.001. https://reader.elsevier.com/reader/sd/F85C22972BB5AADEB AAF06160FBF08F100A0F6C1B728449242FBE8E3FBCFD 1DC1DE573C0993796A873CFF498BE7BB173. Accessed on September 23, 2019.

This somewhat technical article provides a superb review of the development of attribution science and the contributions it has made to our current understanding of human influences on natural disasters.

Fleming, James Rodger. 1998. "Arrhenius and Current Climate Concerns: Continuity or a 100-year Gap?" *Eos* 79(34): 405–410. https://agupubs.onlinelibrary.wiley.com/doi/pdf/10 .1029/98EO00310. Accessed on September 22, 2019.

The author reviews Arrhenius's research on climate change and points how different his views of the impact of climate change would be on humans than is held today.

Glaser, Rüdiger, Iso Himmelsbach, and Annette Bösmeier. 2017. "Climate of Migration? How Climate Triggered Migration from Southwest Germany to North America during the 19th Century." *Climate of the Past* 13: 1573–1592. https:// www.clim-past.net/13/1573/2017/cp-13-1573-2017.pdf. Accessed on September 23, 2019.

One of the interesting fields of climate research has to do with evidence for the effects of climate change on human societies in past history. This article attempts to determine the extent to which climate change during the 19th century influenced migration patterns from Germany to North America. The list of references in this article provide several other papers on the ways in which climate change has affected other aspects of human society in the past.

Griscom, Bronson W., et al. 2017. "Natural Climate Solutions." *Proceedings of the National Academy of Sciences* 114(44): 11645–11650. https://www.pnas.org/content/114/44/11645 #sec-1. Accessed on September 24, 2019.

Authors of this report attempt to assess the relative effectiveness of 20 different solutions for climate change based on natural systems. Among the systems considered are reforestation, fire management, grazing practices, improved rice growing systems, coastal restoration, and peat restoration. They find the most effective systems are reforestation and avoidance of forest conversions.

Hafner, Sarah, Olivia James, and Aled Jones. 2019. "A Scoping Review of Barriers to Investment in Climate Change Solutions." *Sustainability* 11(11): 3201. doi:10.3390/su11113201. https://www.mdpi.com/2071-1050/11/11/3201. Accessed on September 24, 2019.

An important, and sometimes neglected, element in discussions about climate change solutions is the need for economic investment in technologies to deal with the problem. The 31 reports in this book discuss reasons that such investments do not occur as frequently as they should and what must take place if progress is to be made.

Hassol, Susan Joy, et al. 2016. "(Un)Natural Disasters: Communicating Linkages between Extreme Events and Climate Change." World Meteorological Organization. *Bulletin* 65(2): 2–9. https://public.wmo.int/en/resources/bulletin/unnatural -disasters-communicating-linkages-between-extreme-events -and-climate. Accessed on September 23, 2019.

The authors discuss the problems with current efforts to connect individual earthquakes, tsunamis, forest fires, or other natural disasters with climate change. They point out the importance of developing methods of communicating these results both to the general public and to experts in the field.

Hollesen, Jørgen, H. Matthiesen, and Bo Elberling. 2017. "The Impact of Climate Change on an Archaeological Site in the Arctic." *Archaeometry* 59(6): 1175–1189. doi:10.1111/arcm.12319. https://curis.ku.dk/ws/files/186676380/The_Impact

_of_Climate_Change_on_an_Archaeological_Site_in_the
_Arctic.pdf. Accessed on September 27, 2019.

Authors of this report comment on the dangers posed to archaeological artifacts in the Arctic as a result of climate change. Also see on this topic, Anderson et al. 2019.

Irvine, Peter, et al. 2019. "Halving Warming with Idealized Solar Geoengineering Moderates Key Climate Hazards." *Nature Climate Change* 9(4): 295–299. https://www.nature.com /articles/s41558-019-0398-8.epdf. Accessed on September 24, 2019.

The authors model the effects of using solar geoengineering to reduce global warming by half. In contrast to previous studies, they find that use of the technology would bring measurable benefits to people worldwide, with essentially no serious risks for societies.

Joireman, Jeff, Manfred Milinski, and Paul Van Lange. 2018. "Climate Change: What Psychology Can Offer in Terms of Insights and Solutions." *Current Directions in Psychological Science* 27(4): 269–274. https://pdfs.semanticscholar.org/2f70/e5caff 85da29317cfc53aad82a572c0a025b.pdf. Accessed on September 24, 2019.

Any review of climate science and climate change reveals the role of people's beliefs, attitudes, and feelings about the topic, to at least as great an extent as their understanding of the science and technology involved. This article discusses some ways in which principles of psychology can be used to inform discussions about the topic.

Kahn, Matthew E., et al. 2019. "Long-Term Macroeconomic Effects of Climate Change: A Cross-Country Analysis." Federal Reserve Bank of Dallas. Globalization Institute Working Paper 365. https://www.dallasfed.org/~/media/documents/institute /wpapers/2019/0365.pdf. Accessed on September 24, 2019.

The authors examine the economic effects of adopting Paris Accord agreements for dealing with climate change versus lack of action in this field on economic growth to 2100. The conclude that former situation results in a 1.07 percent decreased in gross domestic product worldwide, while the latter results in a 7.22 percent loss in GDP. They point out that economic effects will differ significantly in various parts of the world.

Labbé, Thomas, et al. 2019. "The Longest Homogeneous Series of Grape Harvest Dates, Beaune 1354–2018, and its Significance for the Understanding of Past and Present Climate." *Climate of the Past* 15: 1485–1501. https://www.clim-past.net /15/1485/2019/. Accessed on September 23, 2019.

This article presents a very different, but interesting, method for reconstructing past climate: by examining the dates of grape harvesting in a region of France. Researchers conclude that "warm extremes in the past were outliers, while they have become the norm in the present time."

Laepple, Thomas, T. Münch, and A. M. Dolman. 2017. "Inferring past Climate Variations from Proxies: Separating Climate and Non-climate Variability." *PAGES Magazine* 25(3): 140–141. http://www.pages-igbp.org/download/docs/magazine/2017 -3/PAGESmagazine_2017(3)_140-141.pdf. Accessed on September 23, 2019.

One of the challenges in using proxies to determine earlier climates is confirming that the changes observed are actually global changes and not changes resulting from local events. This article describes how such attributions can be made.

Lamb, William F., et al. 2019. "Learning about Urban Climate Solutions from Case Studies." *Nature Climate Change* 9(4): 279–287.

The risks to urban areas posed by climate change is a topic of considerable interest to researchers. This review article addresses several modifications that have been attempted or proposed for urban areas worldwide in order to arrive at some general principles to be applied to such situations.

Lawrence, Mark G. 2018. "Evaluating Climate Geoengineering Proposals in the Context of the Paris Agreement Temperature Goals." *Nature Communications* 9(1): 1–19. https://www.nature.com/articles/s41467-018-05938-3. Accessed on September 24, 2019.

A variety of technical fixes have been suggested to deal with climate change. The authors of this review assess the ability of any one or combination of these technologies to reduce climate change. They find that none of the technologies would be able to help the world meet the Paris Accords of 2015.

Li, Bo, Douglas W. Nychka, and Caspar M. Ammann. 2010. "The Value of Multiproxy Reconstruction of Past Climate." *Journal of the American Statistical Association* 105(491): 883–895. https://opensky.ucar.edu/islandora/object/articles %3A18119/datastream/PDF/view. Accessed on September 23, 2019.

Basing estimates of previous climate change on a single proxy is a risky proposition. This article explains why and how such predictions can be far more reliable when a combination of proxies is used.

MacDonald, Alexander E. 2001. "The Wild Card in the Climate Change Debate." *Issues in Science and Technology* 17(4): 51–56. https://issues.org/macdonald/. Accessed on September 21, 2019.

The author suggests that the "wild card" missing from the current climate change debate is possible regional effects that differ from effects on a worldwide scale. He argues that such changes "may happen fairly quickly, last for a long time, and bring devastating consequences."

McKenzie, Richard B. 2019. "The Climate-Change Dooms-day Trap." *Regulation* 42(2): 28–33. https://www.cato.org/sites/cato.org/files/serials/files/regulation/2019/6/reg-v42n2-2.pdf. Accessed on September 21, 2019.

> The author argues that climate scientists' "frantic" efforts to warn the world about climate change may be a major factor in bringing to reality some of their worst fears. It could, he said, be a matter of conveying the message that there is "no hope" for avoiding the worst consequences of climate change unless action is taken *now*.

Muri, Helene, et al. 2018. "Climate Response to Aerosol Geo-engineering: A Multimethod Comparison." *Journal of Climate* 31(16): 6319–6340. https://journals.ametsoc.org/doi/full/10.1175/JCLI-D-17-0620.1. Accessed on September 24, 2019.

> The authors compare three methods of geoengineering—stratospheric aerosol injections, marine sky brightening, and cirrus cloud thinning—with regard to their likely effectiveness in reducing climate change. They find essentially no difference among the three, with any one most likely able to meet the lowest standards of climate change reduction.

Murphy, Robert P., Patrick J. Michaels, and Paul C. Knappen-berger. 2016. "The Case against a U.S. Carbon Tax." Policy Analysis. Cato Institute. Number 801. https://www.cato.org/sites/cato.org/files/pubs/pdf/pa801.pdf. Accessed on September 21, 2019.

> This pamphlet is produced by the Cato Institute, an entity that has raised questions and published articles about some problems with the concerns about global warming expressed by many climate scientists. This paper explains why the idea of a carbon tax is not a good idea for the United States.

Obradovich, Nick. 2017. "Climate Change May Speed Democratic Turnover." *Climatic Change* 140(2): 135–147.

The author explores the proposition that climate change may produce effects in some unexpected areas, in this case, the political system adopted by a country. For articles on other examples of unexpected consequences of climate change, see Obradovich et al. 2017; Obradovich, Tingley, and Rahwan 2018; "The Most Unexpected Result of Climate Change" 2019.

Obradovich, Nick, Dustin Tingley, and Iyad Rahwan. 2018. "Effects of Environmental Stressors on Daily Governance." *PNAS* 115(35): 8710–8715. doi:10.1073/pnas.1803765115. https://www.pnas.org/content/115/35/8710. Accessed on September 26, 2019.

Several studies have been conducted recently on "unexpected results" of climate change, that is, changes that might occur above and beyond the better-known effects on oceans, the atmosphere, and other parts of the environment. This study examines the effects of climate change on the behavior of first responders to local and regional emergencies. For other examples of "unexpected results" studies, see Obradovich 2017; Obradovich et al. 2017; "The Most Unexpected Result of Climate Change" 2019.

Oster, Jessica, et al. 2019. "Speleothem Paleoclimatology for the Caribbean, Central America, and North America." *Quaternary* 2(1): 5. https://www.mdpi.com/2571-550X/2/1/5/xml. Accessed on September 22, 2019.

This paper provides an overview of the way in which cave minerals can be used as a proxy for climate change. An excellent general introduction and overview to the topic.

Phalkey, R. K., and V. R. Louis. 2016. "Two Hot to Handle: How Do We Manage the Simultaneous Impacts of Climate Change and Natural Disasters on Human Health?" *European Physical Journal: Special Topics* 225(3): 443–457.

The authors take note of the fact that there is now abundant evidence about the effects of natural disasters and of climate change on public health issues. They also point out the new issues posed by the interaction of these two forces to produce even more serious health problems. They also show that public health issues resulting from these forces are likely to affect developing nations to a much greater extent than developed nations. They discuss the planning issues posed by this combination of factors.

Smerdon, Jason E. 2017. "What Was Earth's Climate Like before We Were Measuring It?" *Significance* 14(1): 24–29. https://rss.onlinelibrary.wiley.com/doi/pdf/10.1111/j.1740 -9713.2017.00999.x. Accessed on September 23, 2019.

Scientists currently make use of a host of techniques—proxies—for measuring ancient climates and climate change. Smerdon provides a nice general introduction to the variety of proxies available to researchers today and the types of results they have produced.

Stults, Missy. 2017. "Integrating Climate Change into Hazard Mitigation Planning: Opportunities and Examples in Practice." *Climate Risk Management* 17: 21–34. https://reader.elsevier .com/reader/sd/28174BC9B80553C281142C1E4943EC655 2E9106A384C614A5BE801E70016EA0940611DF8E98E04 860F70B073ACC1C0CA. Accessed on September 23, 2019.

The author summarizes her research on the extent to which state governments have incorporated FEMA recommendations for including climate change in their disaster management plans. She finds that 23 of 35 states studies have indicated a commitment to FEMA's recommendations, although they have, in general, not yet developed specific actions that would result in specific preparedness programs or actions.

Weart, Spencer. 2003. "The Discovery of Rapid Climate Change." *Physics Today* 56(8): 30–37. https://physicstoday .scitation.org/doi/10.1063/1.1611350. Accessed on September 20, 2019.

One of the features of climate change today that needs to be explained is the speed at which that change appears to be occurring. The author provides a historical review of that issue with a discussion of the speed of climate change today.

Reports

Abram, Nerilie, et al. 2019. "IPCC Special Report on the Ocean and Cryosphere in a Changing Climate." Intergovernmental Panel on Climate Change. https://report.ipcc.ch/srocc /pdf/SROCC_FinalDraft_FullReport.pdf. Accessed on September 25, 2019.

One of the most recent and most sobering reports on the development of climate change was the special report on the oceans and the cryosphere, released in September 2019. It describes changes that have already occurred in Earth's hydrosphere, along with projected changes over the next century. The report also describes some efforts that have been and are already being made to deal with this issue.

"Attribution of Extreme Weather Events in the Context of Climate Change." 2016. National Academies of Sciences, Engineering, and Medicine. Washington, DC: The National Academies Press. doi:10.17226/21852. https://www.nap.edu /download/21852#. Accessed on September 23, 2019.

This report explores the extent to which individual natural disasters can be correlated with climate change. The authors conclude that such connections can now be made with a reasonable level of confidence for some natural disasters, such as heat waves, drought, and heavy precipitation. The evidence for such correlations for other disasters is still less convincing, although suggestive of such connections.

Bedsworth, Louise, et al. 2018. "Statewide Summary Report. California's Fourth Climate Change Assessment." Publication number: SUMCCCA4-2018-013. https://www.energy.ca.gov /sites/default/files/2019-07/Statewide%20Reports-%20SUM -CCCA4-2018-013%20Statewide%20Summary%20Report .pdf. Accessed on September 23, 2019.

As with the federal government, the state of California requires a regular report on the status of climate change science and its impacts on the state. The report available here is the fourth such document in that series.

"Climate Change." 2019. Congressional Budget Office. https://www.cbo.gov/topics/climate-and-environment/climate -change. Accessed on September 23, 2019.

The CBO produces irregular reports on various aspect of climate change, such as the cost of hurricane damage and storm-related flooding as a result of climate change, renewable fuel standards, effects of a carbon tax on the environment and the economy, capturing and storing carbon dioxide, and current and recent legislation on climate change.

"Climate Change." 2019. Economic Research Service. United States Department of Agriculture. https://www.ers.usda .gov/topics/natural-resources-environment/climate-change/. Accessed on September 24, 2019.

The ERS regularly conducts studies on various aspects of the effects of climate change on U.S. agricultural activities. This web page provides an overall summary of the findings of this research, with links to more than a dozen studies on specific topics.

Field, Christopher B., et al. 2012. "Managing the Risks of Extreme Events and Disasters to Advance Climate Change Adaptation. Special Report of the Intergovernmental Panel on Climate Change." Intergovernmental Panel on Climate Change. Cambridge, UK: Cambridge University Press. https:// www.ipcc.ch/pdf/special-reports/srex/SREX_Full_Report.pdf. Accessed on August 3, 2018.

This special report from the IPCC focuses on the relationship between extreme weather events and climate change. Its major sub-divisions deal with "Climate Change: New Dimensions in Disaster Risk, Exposure, Vulnerability, and Resilience," "Determinants of Risk: Exposure and Vulnerability," "Changes in Climate Extremes and their Impacts on the Natural Physical Environment," "Changes in Impacts of Climate Extremes: Human Systems and Ecosystems," "Managing the Risks from Climate Extremes at the Local Level," "National Systems for Managing the Risks from Climate Extremes and Disasters," "Managing the Risks: International Level and Integration across Scales," "Toward a Sustainable and Resilient Future," and "Case Studies."

Fletcher, Luke, et al. 2018. "Beyond the Cycle." Carbon Disclosure Project. https://6fefcbb86e61af1b2fc4-c70d8ead6c ed550b4d987d7c03fcdd1d.ssl.cf3.rackcdn.com/cms/reports /documents/000/003/858/original/CDP_Oil_and_Gas _Executive_Summary_2018.pdf (Executive summary only). Accessed on September 25, 2019.

This agency conducts investor research on major industries, such as the automotive industry, chemicals, and electrical utilities. This study reports on the fossil fuel industry, with particular attention to its efforts to meet requirements of the Paris Accord of 2015.

"Fostering Effective Energy Transition. 2019 Edition." 2019. World Economic Forum. http://www3.weforum.org/docs /WEF_Fostering_Effective_Energy_Transition_2019.pdf. Accessed on September 27, 2019.

Authors of this report note that energy systems are always in a state of transition, but the need to combat climate change has significantly altered the pace at which these transitions have to take today. In this follow-up of earlier reports on the topic, the Forum introduces a measure,

Energy Transition Index (ETI), indicating the progress with which each of nearly 200 countries worldwide have achieved the capability of undertaking necessary energy transitions outlined by the Paris Accord of 2015. Nations with the highest ETI are primarily European countries, Sweden, Switzerland, Norway, Finland, Denmark, Austria, and the United Kingdom. Meanwhile, African and South American countries tend to fall near the end of the list, with the lowest five ETI scores going to Mozambique, Venezuela, Zimbabwe, South Africa, and Haiti.

Grant, Andrew, and Mike Coffin. 2019. "Breaking the Habit." Carbon Tracker. https://www.carbontracker.org/reports/breaking -the-habit/. Accessed on September 25, 2019.

Carbon Tracker is an agency devoted to keeping track of the efforts of important industrial agencies, such as the fossil fuel industry, to see how closely they are following the recommendations of the 2015 Paris Accords and to help markets understand this progress or lack of progress.

"Greening with Jobs." 2018. World Employment Social Outlook 2018. Geneva: International Labour Office. https://www .ilo.org/weso-greening/documents/WESO_Greening_EN _web2.pdf. Accessed on September 25, 2019.

The ILO has been studying the effects of climate change on the prospects for the labor market in coming decades. It concludes that dealing with climate change will be a major force in increasing the job market in most parts of the world. This report focuses on topics such as "Environmental Sustainability and Decent Work," "Employment and the Role of Workers and Employers in a Green Economy," "Regulatory Frameworks: Integration, Partnerships, and Dialogue," "Protecting Workers and the Environment," and "Skills for the Green Transition."

Herring, Stephanie C., et al. 2018. "Explaining Extreme Events of 2016 from a Climate Perspective." Special Supplement to

the *Bulletin of the American Meteorological Society* 99(1). http://
www.ametsoc.net/eee/2016/2016_bams_eee_low_res.pdf.
Accessed on September 23, 2019.

This report is the sixth annual report by the AMS about
the influence of climate change on certain types of natural disasters. Thirty papers in the report examine detailed
and technical aspects of specific events, such as those involving ocean heat waves, forest fires, snow storms and
frost, heavy precipitation, drought, and extreme heat and
cold events over land. Reports from previous years are
https://www.ametsoc.org/ams/index.cfm/publications
/bulletin-of-the-american-meteorological-society-bams
/explaining-extreme-events-from-a-climate-perspective/.
Accessed on August 11, 2018.

Hope, Mat. 2018. "Here is What #ShellKnew about Climate
Change in the 1980s." Resilience. https://www.resilience.org
/stories/2018-04-05/here-is-what-shellknew-about-climate
-change-in-the-1980s/. Accessed on September 23, 2019.

Several studies have shown that fossil fuel companies
were well aware by the 1980s of the risks posed by climate change. Those companies reacted in different ways
to this new information. This article summarizes the response by Shell Internationale to this information. A link
to the company's own report at the time is provided in
the article.

Intergovernmental Panel for Climate Change, various dates,
various titles.

The IPCC has issued five major reports on the state of
climate change, in 1990, 1995, 2001, 2007, and 2014. Its
next summary report is scheduled to be released in 2020.
In addition to these major reports, the organization has
published several other reports on more specific topics,
such as Climate Change and Land, The Ocean and Cryosphere, Global Warming of 1.5°C, Renewable Energy

Sources and Climate Change Mitigation, Safeguarding the Ozone Layer and the Global Climate System, and Carbon Dioxide Capture and Storage. All reports can be accessed at ipcc.ch/reports.

"Investing in a Time of Climate Change—The Sequel." 2019. Mercer. https://www.mercer.com/our-thinking/wealth/climate -change-the-sequel.html#. Accessed on September 23, 2019.
Mercer is an international consulting company interested in a wide variety of fields, including healthcare, workforce and careers, investment, and consulting and advising. This report is a follow-up on an earlier report on the directions of climate change and implications for economic decisions within the context of these changes.

Jacobs, R. P. W. M., et al. 1988. "The Greenhouse Effect." Shell Internationale. Report Series HSE 88-001. https://www .documentcloud.org/documents/4411090-Document3.html. Accessed on September 23, 2019.
Increasing information about possible climate changes was becoming available to the world in the late 1960s and 1970s. One consequence of this growing awareness by researchers was that, by the 1980s, most fossil fuel companies were beginning to examine this information and assess the role that their own activities might be playing in climate change. Most also commissioned formal reports from their own scientists on the best response their companies could make. This document is a reproduction of this type of endeavor by one of the world's largest oil companies. The company's position on its role in climate change has evolved over time. Its current views on climate change and Shell's actions on the issue can be found in "Shell Energy Transition Report" at https://www.shell.com/energy-and-innovation/the -energy-future/shell-energy-transition-report/_jcr_content /par/toptasks.stream/1524757699226/3f2ad7f01e2181c3 02cdc453c5642c77acb48ca3/web-shell-energy-transition -report.pdf. Accessed on September 23, 2019.

National Audubon Society. 2015. "Audubon's Birds and Climate Change Report: A Primer for Practitioners." New York: National Audubon Society. http://climate.audubon.org/sites /default/files/NAS_EXTBIRD_V1.3_9.2.15%20lb.pdf.   Accessed on September 23, 2019.

This report summarizes research on the likely effect of climate change on 588 species of birds present in the country. It uses three different models of climate change to obtain their predictions. Just over half (314 species; 53.4 percent) lost more than half of their geographic range as a result of climate change. Nearly half of that number (126 species) are unable to expand their range to make up for this loss, while 188 species appear to be capable of achieving that result.

"Psychology and Global Climate Change: Addressing a Multi-faceted Phenomenon and Set of Challenges." 2011. American Psychology Association. https://www.apa.org/science /about/publications/climate-change-booklet.pdf. Accessed on September 27, 2019. This report also appeared in *American Psychologist* 66(4): 241–328.

This report reflects the fact that discussions over climate change also bring with them many elements of psychological theory and practice. The main topics considered in the report are how people understand the risks posed by climate change, how human behavior contributes to climate change, what the psychosocial impacts of climate change are on an individual, how people adapt to the perceived threats posed by climate change, what psychological barriers contribute to or limit action on climate change, and how psychologists can help limit climate change.

"Report." 2019. Committee on Climate Change. https://www .theccc.org.uk/publicationtype/0-report/. Accessed on September 23, 2019.

The Committee on Climate Change was created by the British Climate Change Act of 2008. Over the past decade, it has produced reports on topics such as The Future of Carbon Pricing in the UK, Reducing UK Emissions, Progress in Preparing for Climate Change, Net Zero, UK Housing: Fit for the Future?, Reducing Emissions in Northern Ireland, and Hydrogen in a Low-Carbon Economy.

Siegmund, Peter, et al. 2019. "The Global Climate in 2015–2019." World Meteorological Organization. https://library.wmo.int/doc_num.php?explnum_id=9936. Accessed on September 25, 2019.

This report on the status of climate change is divided into eight parts: Greenhouse Gases, Temperature, Ocean, Cryosphere, Precipitation, Extreme Events, Attribution of Extreme Events, and Highlights on Prominent Climate-related Risks. The report is highlighted by several helpful graphs, charts, and illustrations.

"Understanding the Climate System." 2019. The National Academies. http://sites.nationalacademies.org/sites/climate/SITES_193558. Accessed on September 23, 2019.

The National Academies has published several reports on the overall effects of climate change on the United States, as well as reports on specific topics, such as Review of Drift Climate Science, Antarctic Sea Ice Variability, Attribution of Extreme Weather Events, The Future of Atmospheric Chemistry Research, and Lessons and Legacies of the International Polar Year, 2007–2008.

United States Global Change Research Program. 2017. "Fourth National Climate Assessment, Volume I. Climate Science Special Report." Washington, DC: Global Change Research Program. doi:10.7930/J0J964J6. https://science2017.globalchange.gov/downloads/CSSR2017_FullReport.pdf. Accessed on September 23, 2019.

The Global Change Research Act of 1990 mandates that the U.S. Global Change Research Program deliver to the president at least every four years a report that "1) integrates, evaluates, and interprets the findings of the Program . . .; 2) analyzes the effects of global change on the natural environment, agriculture, energy production and use, land and water resources, transportation, human health and welfare, human social systems, and biological diversity; and 3) analyzes current trends in global change, both human-induced and natural, and projects major trends for the subsequent 25 to 100 years." (This report, page 1) The Fourth National Climate Assessment report, NCA4, appeared in two parts, the one listed here and the following item. The report provides a detailed summary of all that is currently known about climate change and its effects on many aspects of American society.

United States Global Change Research Program. 2018. "Fourth National Climate Assessment. Volume II. Impacts, Risks, and Adaptations in the United States." Washington, DC: Global Change Research Program. doi:10.7930/NCA4.2018. https://nca2018.globalchange.gov/downloads/NCA4_2018_Full Report.pdf. Accessed on September 23, 2019.

See previous listing.

Internet

Alderman, Liz. 2019. "What Worries Iceland? A World without Ice. It Is Preparing." *New York Times.* https://www.nytimes.com/2019/08/09/business/iceland-ice-melt-global-warming -climate-change.html. Accessed on September 24, 2019.

One of the countries likely to be most severely affected by climate change in the short term is Iceland. This article explains how the nation has already been affected by climate change and what it is doing to respond to future climate alterations.

"Attribution of the 2018 Heat in Northern Europe." 2018. World Weather Attribution. https://www.worldweatherattribution

.org/analyses/attribution-of-the-2018-heat-in-northern -europe/. Accessed on September 23, 2019.

This study, done in real time analysis of weather patterns in northern Europe in 2018, The study concludes that "the probability of such a heatwave to occur has increased everywhere in this region due to anthropogenic climate change," and that "the probability to have such a heat or higher is generally more than two times higher today than if human activities had not altered climate.

Berardelli, Jeff. 2019. "How Climate Change Is Making Hurricanes More Dangerous." Yale Climate Connections. https:// www.yaleclimateconnections.org/2019/07/how-climate -change-is-making-hurricanes-more-dangerous/. Accessed on September 24, 2019.

A question that arises frequently is how climate change is likely to affect severe storms. The answer appears to be that such storms will not necessarily occur more frequently, but they are like to be substantially more severe. This article discusses this trend.

"Can an Increase or Decrease in Sunspot Activity Affect the Earth's Climate?" n.d. National Weather Service. https://www .weather.gov/fsd/sunspots. Accessed on September 20, 2019.

Many discussions of climate change ignore the role played by the Sun. This website provides useful background information on the topic.

"Carbon Pricing 101." 2018. Union of Concerned Scientists. https://www.ucsusa.org/global-warming/reduce-emissions/cap -trade-carbon-tax. Accessed on September 20, 2019.

The three main sections of this article deal with science considerations, equity concerns, and carbon pricing in action. A good general introduction to the topic.

Carbon Tax Center. 2018. Carbon Tax. https://www.carbontax .org/. Accessed on September 20, 2019.

This website is an excellent resource for learning about all aspects of carbon taxes, including an explanation of what a carbon tax is, where it has been implemented, why it is a promising solution for the issue of climate change, what the status of carbon taxes in China is, and related issues.

Cho, Renee. 2019. "How Climate Change Impacts the Economy." State of the Planet. https://blogs.ei.columbia.edu/2019 /06/20/climate-change-economy-impacts/. Accessed on September 20, 2019.

This article provides a good summary of the various aspects of the economy that are affected by climate change, and what those affects are likely to be.

"Climate Change." 2019. The Heartland Institute. https:// www.heartland.org/topics/climate-change/. Accessed on September 21, 2019.

The Heartland Institute is a well-known climate change skeptic/denier institution. This website presents its position on several climate change related issues, such as biological effects of global warming, economic effects of climate change, and the overall significance of any climate change that may be occurring.

"Climate Change: How Do We Know?" 2019. Global Climate Change. NASA. https://climate.nasa.gov/evidence/. Accessed on September 20, 2019.

This website provides one of the most complete and easily understood collections of facts about climate change, with links to sources with more detailed descriptions of each category of facts.

"Courses about Climatology and Climate Change." 2016. Teach the Earth. https://serc.carleton.edu/NAGTWorkshops /climatechange/courses.html. Accessed on September 27, 2019.

This website provides links to 30 courses teachers have developed for teaching about climate change.

Darby, Megan. 2019. "Net Zero: The Story of the Target That Will Shape Our Future." Climate Home News. https://www .climatechangenews.com/2019/09/16/net-zero-story-target -will-shape-future/. Accessed on September 26, 2019.

A growing goal for climate change control is called *net zero*. It is based on the effort to ensure that at some point in the future, the total amount of carbon dioxide released to the atmosphere will be no more nor less than the amount removed from the atmosphere. This article provides a good general overview of the movement, with a discussion of the rationale and history of its origins.

Denbow, J. 2012. "Palynology." University of Texas. https:// la.utexas.edu/users/denbow/labs/palynology.htm. Accessed on September 20, 2019.

This website provides a clear description of the process of palynology, with references to some specific research in the field.

Evers-Hillstrom, Karl, and Raymond Arke. 2019. "Fossil Fuel Companies Lobby Congress on Their Own Solutions to Curb Climate Change." OpenSecrets.org. https://www.opensecrets .org/news/2019/05/fossil-fuel-lobby-congress-on-climate -change/. Accessed on September 23, 2019.

Over the past few decades, fossil fuel companies have altered their views on the existence of climate change, the role of human activities in that change, and the responses they, the companies, should make to future climate issues. This article provides a superb review of the current (2019) views of fossil fuel companies on these issues. It contains many useful links to specific aspects of the report.

Gattuso, Jean-Pierre, et al. 2018. "Ocean Solutions to Address Climate Change and Its Effects on Marine Ecosystems."

Frontiers in Marine Science 5. doi:10.3389/fmars.2018.00337.
https://www.frontiersin.org/articles/10.3389/fmars.2018
.00337/full. Accessed on September 24, 2019.

In this review article, researchers review 13 suggested
solutions for climate change based on ocean-focused
programs.

"Geoengineering Monitor." 2019. http://www.geoengineering
monitor.org/. Accessed on September 24, 2019.

This website is maintained by an organization opposed
to the development and employment of geoengineering
technologies for the control of climate change. The four
main reasons for their opposition are that the technologies
don't work; they can be converted to weapons systems;
they detract from "real solutions" to the climate change
problem; and they damage human rights and biodiversity.

Hafstead, Marc. 2019. "Carbon Pricing 101." Resources for
the Future. https://media.rff.org/documents/Carbon_Pricing
_Explainer.pdf. Accessed on September 20, 2019.

This article provides an introduction to carbon taxes and
cap-and-trade programs, the benefits and design of each,
and some ways they have been adopted in nations around
the globe.

Hasemyer, David. 2019. "Fossil Fuels on Trial: Where the
Major Climate Change Lawsuits Stand Today." Inside Climate
News. https://insideclimatenews.org/news/04042018/climate
-change-fossil-fuel-company-lawsuits-timeline-exxon-children
-california-cities-attorney-general. Accessed on September 23,
2019.

Over the past decade, information has become available about what fossil fuel companies knew about climate change and their own involvement in the process
of global warming. As this information unfolds, governmental agencies and non-profit groups have begun to file

suits against companies for their disregard of their role in bringing about climate change. This article provides an excellent summary of this situation, along with a valuable timeline of the status of existing suits against the companies.

Hausfather, Zeke. 2019. "Factcheck: What Greenland Ice Cores Say about Past and Present Climate Change." Carbon-Brief. https://www.carbonbrief.org/factcheck-what-greenland -ice-cores-say-about-past-and-present-climate-change.Accessed on September 23, 2019.

A dispute has raged in the academic world over the past 25 years about the significance of proxy data from Greenland about climate change on the globe over the past two millennia. This paper attempts to resolve that debate, pointing out problematic issues with earlier research. Helpful graphs explain the findings presented in this report.

"Heavy Weather: Tracking the Fingerprints of Climate Change, Two Years after the Paris Summit." 2017. Energy & Climate Intelligence Unit. https://issuu.com/eciu/docs/eciu_climate _attribution_report_dec. Accessed on December 14, 2019.

In recent years, researchers have become increasingly interested in efforts to connect specific disaster events with climate change. This article summarizes the results of a search for articles of this type in the two-year period following the Paris summit on climate change in 2015.

"Highlights: The Net-zero Climate Change Conference in Oxford." CarbonBrief. https://www.carbonbrief.org/highlights -the-net-zero-climate-change-conference-oxford. Accessed on September 27, 2019.

The United Kingdom became the first nation in the world in 2019 in specifically adopting a "net zero" policy for the reduction of carbon emissions as a tool for fighting climate change. The conference reported here was designed as a general, national introduction to the concept of net zero, the country's current status with regard to carbon emissions, and directions for the future in this effort. The article contains numerous links to specific papers presented at the conference.

Hoffman, Andrew J. 2012. "Climate Science as Culture War." Stanford Social Innovation Review. https://ssir.org/book _reviews/entry/climate_science_as_culture_war#. Accessed on September 24, 2019.

The author argues that the world's inability to find solutions for climate change is less a matter of science and technology and more of fundamental differences in social and political philosophies. The article is accompanied by a long list of responses to and commentaries about this analysis.

Kaminski, Isabella. 2019. "In Courtrooms, Climate Change Is No Longer Up for Debate." Undark. https://undark.org/article /in-courtrooms-climate-change-is-no-longer-up-for-debate/. Accessed on September 21, 2019.

The author provides evidence that the reality of global warming and climate change is no longer a matter of dispute in courts. "Increasingly," she says, "global warming science is going unchallenged in lawsuits seeking to curb fossil fuel use and hold companies to account." She explains why this is and what its legal consequences may be.

Lomborg, Bjørn. 2019. "The Danger of Climate Doomsayers." Project Syndicate. https://www.project-syndicate.org /commentary/climate-change-fear-wrong-policies-by-bjorn -lomborg-2019-08. Accessed on September 26, 2019.

Lomborg is generally regarded as one of the most outspoken and well-informed climate change skeptics in the world. In this article, he lays out the major reasons for his beliefs that, during a period of global warming, "the world is mostly getting better."

Mason, John. 2013. "The History of Climate Change." Skeptical Science. https://skepticalscience.com/history-climate -science.html. Accessed on September 20, 2019.

> Among the many histories of climate change, this website provides one of the most complete and well written options.

"Media Reaction: Amazon Fires and Climate Change." 2019. CarbonBrief. https://www.carbonbrief.org/media-reaction-am azon-fires-and-climate-change. Accessed on September 24, 2019.

> The Amazon forest plays a critical part in many discussions about climate change. The region has sometimes been called the planets "lungs" because of their role in absorbing carbon dioxide from the atmosphere. Widespread fires in the region in 2019 further raised questions about the causes of such fires, their possible control, and their effects on the region and the planet. This web page brings together a large collection of articles about the topic.

Metcalf, Gilbert. 2019. "Carbon Taxes: What Can We Learn from International Experience?" Econofact. https://econofact .org/carbon-taxes-what-can-we-learn-from-international -experience. Accessed on September 20, 2019.

> The author draws on his own research on the economic effects of climate change on the economy of British Columbia to draw some general conclusion as to what that scenario might be like worldwide in the future.

Moore, Johnnie N. 2014. "Death by Degrees." Clima Nova. https://climanova.wordpress.com/2014/10/29/death-by -degrees/. Accessed on September 20, 2019.

> During the late 1960s and 1970s, many researchers became concerned about the onset of a "new ice age." Data over that time suggested that such a possibility existed. This article reviews the history of that period and

illustrates how it was eventually replaced by a concern about global warming.

"The Most Unexpected Effect of Climate Change." 2019. Inter-American Development Bank. https://www.iadb.org/en /improvinglives/most-unexpected-effect-climate-change.   Accessed on September 26, 2019.

The effects of climate change on global temperatures, composition of the atmosphere, sea levels, glaciers and ice sheets, agriculture, food security, and other fields have been studied extensively. Less often mentioned are consequences of climate change that might not be expected or even thought of. This article discusses one such event, a dramatic decrease in the production of coffee beans at a time when the demand for coffee is growing substantially. It reviews the reasons for such a change and actions that can be taken to adapt to changing global environments. (For other examples of "unexpected results" from climate change, also see Obradovich 2017; Obradovich, et al. 2017; Obradovich, Tingley, and Rahwan 2018.)

"Nestlé Accelerates Action to Tackle Climate Change and Commits to Zero Net Emissions by 2050." 2019. https://www .nestle.com/sites/default/files/2019-09/press-release-climate -change-zero-net-emissions-2050-en.pdf. Accessed on September 27, 2019.

In 2019, the Nestlé corporation announced that it would be one of the first major industries in the world to specifically adopt a net zero production policy for its operations. This press release describes some of the specific actions the company as a way of reaching this goal.

"Net Zero: Why?" 2018. Energy & Climate Intelligence Unit. https://ca1-eci.edcdn.com/briefings-documents/net-zero-why -PDF-compressed.pdf. Accessed on September 27, 2019.

This briefing outlines the principle of net zero emissions as a way of ameliorating climate change, the science of carbon budgets, and the progress of specific nations in moving toward net zero emissions. It pays special attention to the policies of the United Kingdom in this field.

"Options and Considerations for a Federal Carbon Tax." 2013. Center for Climate and Energy Solutions. https://www .c2es.org/site/assets/uploads/2013/02/options-considerations -federal-carbon-tax.pdf. Accessed on September 20, 2019.

This summary is a somewhat detailed description of carbon pricing, with information on carbon taxes in a variety of countries.

PAGES2k Consortium. 2017. "A Global Multiproxy Database for Temperature Reconstructions of the Common Era." Nature .com. https://www.nature.com/articles/sdata201788.pdf. Accessed on September 23, 2019.

Many articles are available using a climate proxy such as ice cores or dendrochronology to determine the age of various historic events and trends. This article takes a somewhat different approach in combining data from all available proxies to estimate temperature changes for the Common Era, year 0 to the present day. Their results produce a graph that looks similar to the famous "hockey stick" graph previously presented by a variety of authors.

"Paleoclimatology Datasets." n.d. National Centers for Environmental Information. https://www.ncdc.noaa.gov/data -access/paleoclimatology-data/datasets. Accessed on September 20, 2019.

This website provides a superb introduction to climate change proxies, such as boreholes, corals, ice cores, insect remains, pollen, and tree rings. Useful links to more detailed information are also available.

Pearce, Fred. 2019. "Geoengineer the Planet? More Scientists Now Say It Must Be an Option." YaleEnvironment360. https://

e360.yale.edu/features/geoengineer-the-planet-more-scientists -now-say-it-must-be-an-option. Accessed on September 24, 2019.

The author sees geoengineering for climate change control as being very popular among scientists. He describes the technological procedures involved and outlines some risks and benefits in the use of the technologies.

Polansky, Anne. 2016. "ExxonMobil and Climate Change: A Story of Denial, Delay, and Delusion, Told in Forms 10-K (1993–2000)." Climate Science & Policy Watch. http:// www.climatesciencewatch.org/2016/03/08/exxonmobil-and -climate-change-a-story-of-denial-delay-and-delusion-told-in -forms-10-k-1993-2000/. Accessed on September 19, 2019.

This series of three articles traces in great detail the approach that oil giant ExxonMobil has taken with regard to climate change over a decade. The article includes a detailed timeline of events that occurred during that period of time.

Popovich, Nadja, Livia Albeck-Ripka, and Kendra Pierre-Louis. 2019. "85 Environmental Rules Being Rolled Back under Trump." *New York Times*. https://www.nytimes.com/interactive /2019/climate/trump-environment-rollbacks.html. Accessed on September 24, 2019.

As noted in Chapter 2 of this book, changes in presidential administrations often bring changes in policies on important issues of the day. This article summarizes some of the changes in climate policy in the transition from that of President Barack Obama to that of Donald Trump.

"The Psychology of Climate Change Communication." 2009. Center for Research on Environmental Decisions. http://guide .cred.columbia.edu/guide/intro.html. Accessed on September 27, 2019.

One of the fundamental issues involved in dealing with climate change is translating scientific information that individuals should and may want to better understand. This online guide consists of eight chapters that lay out strategies through which individuals can discuss the meaning and significance of climate change.

Ridley, Matt. 2013. "Why Climate Change Is Good for the World." The Spectator. https://www.spectator.co.uk/2013/10/carry-on-warming/. Accessed on September 26, 2019.

The author makes his case in the sub-title of the article: "Don't panic! The scientific consensus is that warmer temperatures do more good than harm." He then quotes professionals in the field who argue that global warming will produce greater benefits than risks for the world.

Riebeek, Holli. 2005. "Paleoclimatology: Introduction." Earth Observatory. NASA. https://earthobservatory.nasa.gov/features/Paleoclimatology_Speleothems. Accessed on September 22, 2019.

Speleothems are mineral formations that form in caves. They can often be used as proxies in climate change studies. This article provides a general introduction to the study of speleothems in research on climate change.

Smith, Warren Cole. 2019. "What if Climate Change Is a Good Thing?" The Stream. https://stream.org/what-if-climate-change-is-a-good-thing/. Accessed on September 26, 2019.

The author explores the possibility that climate change may have its advantages in a future world, such as a decrease in the rate of population growth. He also points out the world has already started to adapt to possible higher global temperatures and that current warnings about the risks of climate change may be exaggerated.

"Teaching Climate." 2019. Climate.gov. NOAA. https://www.climate.gov/teaching#slideshow-1. Accessed on September 27, 2019.

Individuals interested in teaching more about climate change in their classes may need leads to resources that can be used in this endeavor. This website provides a good general overview to the resources that are available, with useful links to specific examples of climate change classes and lessons.

Turner, James Morton, and Andrew C. Isenberg. 2018. "The Climate Change Debate No Longer Turns on Science. That May Not Be All Bad." WBUR. https://www.wbur.org/cognoscenti /2018/12/19/trump-global-warming-james-morton-turner -and-andrew-c-isenberg. Accessed on September 21, 2019.
    The writers argue that the scientific information about climate change is now relatively well known by the general public. The final decisions as to what to do—and not do—they suggest will be based on value judgments, not scientific data.

Weart, Spencer. 2019. "The Discovery of Global Warming." American Institute of Physics. https://history.aip.org/climate /index.htm. Accessed on September 20, 2019.
    This website is far and away the best single source of information about the history of global warming and climate change. It includes articles on virtually every possible aspect of the topic, with links to a great many primary sources. An absolute must for anyone interested in the history of climate science.

Weart, Spencer Weart, and Raymond T. Pierrehumbert. 2007. "A Saturated Gassy Argument." RealClimate. http://www .realclimate.org/index.php/archives/2007/06/a-saturated-gassy -argument/. Accessed on September 20, 2019.
    This two-part article provides a very interesting discussion of the discovery and early research on the greenhouse effect, as well as the controversies that developed over the findings and interpretations resulting from that research.

Introduction

Earth's climate has been changed since the planet was formed some 4.5 billion years ago. Paleoclimatologists have developed a sophisticated set of tools for discovering the nature of those changes up until the time when modern measuring devices were available, about 1850. After that time, more and better records were available about the nature of changes in the planet's climate. As these data became available, researchers also began to develop more detailed theories as to how these changes come about, especially, in the past century, how human activity has contributed to climate change. This chapter lists some of the most important events in that millennial-long history of global climate change.

ca. 2.5 million years B.P.   Ice caps first appear on Earth. Scientists believe that Earth's annual temperature was too high (16° to 23°C [60° to 73°F]) for permanent ice fields to form.

ca. 18,000 years B.P.   The most recent ice age reaches its peak and begins to decrease in area, volume, and extent.

ca. 11,000 B.P.   The current interglacial period is thought to have begun at about this time.

The Giant Hands sculpture by Lorenzo Quinn appears to hold up a building along the Grand Canal in Venice, Italy. The sculpture shows concern for the effects of climate change. (Crackerclips/Dreamstime.com)

ca. 1300 C.E.   An estimated date at which the so-called Little Ice Age began. The period lasted until about 1700 C.E.

ca. 1645–1715   Range of the Maunder Minimum, thought to have been the coldest period of the Little Ice Age.

1712   English inventor Thomas Newcomen invents the steam engine. But see **1769**.

ca. 1750 C.E.   Ice core studies indicate that the concentration of carbon dioxide in the atmosphere reaches about 280 parts per million (ppm). Over the next century, that number increases by about 30 ppm.

1753   Scottish chemist Joseph Black discovers carbon dioxide by treating limestone (calcium carbonate) with acid.

1769   Scottish inventor James Watt significantly improves the steam engine, originally developed by Thomas Newcomen in **1712**. Watt's work is often said to be the true origin of the modern steam engine.

1827   French mathematician Jean-Baptiste Fourier outlines a process by which solar energy is captured by Earth's atmosphere, thus raising the planet's temperature. He is sometimes said to have invented the term *greenhouse effect*, although there appears to be no mention of the term in any of his published writings. See, then, **1909**.

1856   American amateur researcher Eunice Foote writes a paper on the role of carbon dioxide in the capture of solar energy in Earth's atmosphere. (Also see **1861**.)

1861   Irish physicist John Tyndall publishes the first of a series of papers on the role of carbon dioxide in global warming. Tyndall's research in the field is more extensive and better known in today's world than is Foote's. (**1856**)

1896   Swedish physicist Svante Arrhenius concludes that carbon dioxide in the atmosphere is a more important factor in global warming than is water vapor, reversing a position held by many researchers of the time. He calculates that doubling the concentration of carbon dioxide in the atmosphere will increase temperatures by 5° to 6°C. He views this prediction as a good omen for the human race because it will greatly reduce the chance of another ice age's forming on Earth.

1900   Swedish physicist Knut Ångstrom conducts a series of experiments on the warming effects of gaseous components of the atmosphere on global temperatures. He reaches a series of incorrect conclusions, most importantly, that carbon dioxide is less important than water vapor or ozone in absorbing heat in the atmosphere. He also argues that the spectrum of carbon dioxide in the atmosphere is already saturated, so the addition of more carbon dioxide will have no effect on global temperatures. His results are in stark contrast with those of Arrhenius, producing a long-standing and contentious debate between the two scientists.

Ångstrom is not particularly concerned about his results since he imagines that the doubling of carbon dioxide as a result of human activities is unlikely to occur for hundreds of years.

1909   The first definite use of the term *greenhouse effect* in print seems to appear in an article by English physicist John Henry Poynting.

1909   American astronomer A. E. Douglass publishes a paper, "Weather Cycles in the Growth of Big Trees," often said to be the seminal paper in the science of dendrochronology, a work for which he is often called the Father of Dendrochronology.

1912   A brief article, "Coal Consumption Affecting Climate," in the August 14 edition of three Australian newspapers, *The Braidwood Dispatch and Mining Journal, The Rodney and Otamatea Times*, and *The Waitemata and Kaipara Gazette* is thought by some authorities to be the first public statement of the relationship between fossil fuel use and global warming.

1920s   Serbian geophysicist and astronomer Milutin Milanković publishes a series of papers describing changes in the amount of solar radiation reaching Earth as the result of three astronomical cycles.

1922   British mathematician and physicist Lewis Fry Richardson develops the earliest form of mathematical modeling of atmospheric circulation. He divides the planet into cells, for each of which he can express temperature, air pressure, and other variables. He is then able to calculate by hand changes that occur between adjacent cells and thus produce predictions for circulation across all cells. He acknowledges that this approach to modeling of the atmosphere is primitive and very complex because the atmosphere itself is complex. The system does not become practical until the development of computer modeling in the 1960s. (See **1967**.)

1922   U.S. consul to Norway George Nicolas Ifft sends a report to the U.S. Weather Bureau about temperature changes in the Arctic. He starts his report with the observation that "[t]he Arctic seems to be warming up." The report later occurs in the bureaus *Monthly Weather Review*. It is sometimes said to be the first reliable indication that climate change has begun to occur in that part of the planet.

1938   English steam engineer Guy Stewart Callendar attempts to show how carbon dioxide released by human activities affects atmospheric temperature.

1948   Belgian American spectroscopist Marcel V. Migeotte announces the discovery of methane in the atmosphere. Methane is later recognized as one of the most important gases associated with heating of the atmosphere.

1956   American meteorologist Norman A. Phillips develops the first computer model for predicting climatic conditions in a region.

1957   Roger Revelle and Hans Suess of Scripps Institute of Oceanography warn in an article in the journal *Tellus* that excess emissions of carbon dioxide are *not* being absorbed by the oceans, as many previous researchers had expected.

1958   Postdoctoral student Charles David Keeling begins a series of observations at Mauna Loa, Hawaii, on the concentration of carbon dioxide in the atmosphere. Those studies continue to the present day.

1960s Many climate scientists begin to speculate about the beginnings of a "new ice age" for Earth. They base their predictions on records from the preceding two decades showing decreases in Earth' annual global temperatures. These dips in temperature eventually turn out to be transitory trends, changing to regular increases in global temperatures ever since that period. Also see **1972** letter to President Richard Nixon on the topic.

1963 The National Oceanic and Atmospheric Administration establishes the Geophysical Fluid Dynamics Laboratory at Princeton University, under the direction of Joseph Smagorinsky. The purpose of the laboratory is to develop mathematical models of the atmosphere for climate analysis and other purpose.

1963 The U.S. Conservation Foundation (now the World Wildlife Foundation) sponsors possibly the first conference dealing with the effects of climate change.

1964 American meteorologist William D. Sellers publishes the world's first textbook on climatology, *Physical Climatology*.

1965 In a report, "Restoring the Quality of Our Environment," President Lyndon B. Johnson's Science Advisory Committee warns that increases in anthropogenic carbon dioxide "may be sufficient to produce measurable and perhaps marked changes in climate, and will almost certainly cause significant changes in the temperature and other properties of the stratosphere."

1967 Japanese American meteorologist and climatologist Syukuro Manabe and American meteorologist Richard T. Wetherald publish a paper that, for the first time, represents all of the basic elements of Earth's climate in a mathematical model, from which predictions about future climate changes could be made.

1968 Russian meteorologist M. I. Budyko explains how small changes in global temperatures can have produced the ice ages.

1969 Weather satellite Nimbus-3 collects very large amounts of data about the presence of greenhouse gases in the atmosphere and atmospheric temperatures at various altitudes. These data provide far more accurate information about global temperatures than had been available by any previous method. Other weather satellites also collect simple atmospheric data useful in predicting possible climate changes.

1972 American geologists George Kukla and Robert Matthews write a letter to President Richard Nixon warning him that dramatic climatic changes might already have begun, changes that would thrust the planet into a new ice age.

1975 American geophysicist Wallace Smith Broecker publishes a paper, "Climatic Change: Are We on the Brink of a Pronounced Global Warming?", that is sometimes regarded as one of the first predictions of serous climate change in coming years.

1979 The World Meteorological Organization sponsors the First World Climate Congress. Similar meetings are continued annually to the present day.

1980 Indian American researcher Veerabhadran ("Ram") Ramanathan finds that trace gases in the atmosphere, such as nitrous oxide, ozone, and CFCs, may contribute as much as 40 percent of all observed global warming, with the remaining 60 percent caused by carbon dioxide.

1980 The Environmental Protection Agency announces the first cap-and-trade plan in the world. It was designed for the control of acid rain, a goal in which it was very successful.

1985 The United Nations Environment Programme, World Meteorological Organization, and International Council for Science conduct a conference held in Villach, Austria, to review current knowledge about carbon dioxide in the atmosphere and its effects on global climate.

1988   American climatologist James Hansen testifies before the U.S. Senate Energy and Natural Resources Committee on the topic of climate change. He reports that studies conducted at his laboratory have produced, with 99 percent certainty, results that prove global warming has actually begun and that the release of carbon dioxide and other greenhouse gases are responsible for this change. The testimony receives widespread attention and is generally thought to mark a turning point in knowledge of and attitudes toward climate change in the United States and other parts of the world.

1988   The United Nations Environment Programme and the World Meteorological Organization create the Intergovernmental Panel on Climate Change (IPCC). The organization is created to provide the world's governments with regular updates on the status of climate research, along with recommendations for actions dealing with global warming and other related changes.

1989   A group of more than 40 energy-related companies and trade associations organize the Global Climate Coalition (GCC) for the purpose of presenting alternative views about the accuracy and relevance of statements about global warming and climate change.

1990   The IPCC issues its first annual report on climate change, First IPCC Assessment Report (FAR).

1990   Finland adopts the world's first carbon tax.

1991   The National Coal Association, Western Fuels Association, and Edison Electrical Institute combine to form the Information Council on the Environment, designed to advocate for the position that global warming is a theory, not a fact.

1992   The United Nations Framework Convention on Climate Change is adopted and signed by 165 nations at the Earth Summit conference, held in Rio de Janeiro, Brazil. The treaty calls for annual meetings of participants of the treaty.

1992   Former vice president Al Gore publishes a popular book, *Earth in the Balance*, in which he summarizes existing evidence about global warming and its future effects on climate change. (Also see **2006**.)

1993   President Bill Clinton announces the Climate Change Action Plan. A major goal of the plan was to return greenhouse gas emissions to their 1990 levels by the year 2000. The president acknowledged that this was an "ambitious" goal, and virtually no progress was ever made in that direction.

1996   The IPCC issues its second annual report on climate change, Second IPCC Assessment Report (SAR).

1996   British Petroleum withdraws from the GCC (see **1989**). Over the next five years, many other important members of the group withdraw from the organization.

1997   The Kyoto Protocol is adopted at the third annual session of the Conference of Parties to the United Nations Framework Convention on Climate Change (UNFCCC), created in Rio in 1992. Andorra, Palestine, and South Sudan were the only entities not to sign the protocol. The United States signed, but never ratified, the treaty. Canada signed and ratified the treaty, but later withdrew from the protocol.

1998   Soviet researchers at the Vostok research station in East Antarctica achieve the deepest ice core ever obtained by humans, a length of 3,623 meters (11,886 feet). That record remains today.

1998   American climatologists Michael E. Mann, Raymond S. Bradley, and Malcolm K. Hughes report on their studies of global temperatures over a period of about 1,000 years. The graph summarizing their findings shows a relatively low, constant value for most of that period, but a sharp upward trend beginning about the end of the 19th century. They cite this data as evidence of human involvement in global warming. The graph is later given the name of *the hockey-stick curve* because of its shape, mimicking that of a hockey stick. Later objections are raised to these findings, but a substantial amount of research eventually confirms the general findings represented by the curve.

2001   The IPCC issues its third annual report on climate change, Third Assessment Report (TAR).

2004+   A series of studies on climatologists' views on climate change finds that about 97 percent of researchers in the field acknowledge that climate change is occurring and that human activities constitute a major cause of that trend.

2005   The European Union initiates the world's first cap-and-trade program for the control of climate change.

2006   Former vice president Al Gore releases a documentary film about the future of climate change, "An Inconvenient Truth." A sequel to the film, "An Inconvenient Sequel: Truth to Power," is released in 2017. The film was awarded the 2016 Academy Award for Best Documentary Feature.

2007   Former vice president Al Gore is awarded a share (along with IPCC) of the Nobel Peace Prize for his efforts on behalf of efforts to educate the general public about the issues of global warming and climate change.

2007   The IPCC issues its fourth annual report on climate change, Fourth Assessment Report (AR4).

2009   A server at the Climatic Research Unit at the University of East Anglia is hacked and thousands of emails and computer files stolen. Some climate change deniers and skeptics used the documents obtained to confirm their view that global warming is not a real phenomenon. The event is sometimes referred to as "climategate." No reliable evidence exists to support this interpretation of the global warming phenomenon.

2013   President Barack Obama (hereinafter, Obama) issues a Climate Action Plan designed to guide the nation's efforts to deal with climate change over coming decades.

2013   Obama announces the Better Buildings Challenge intended to promote improved efficiency of energy use in commercial and industrial facilities.

2014   Obama creates the Quadennial Energy Review, for the purpose of establishing a comprehensive strategy for finding and implementing climate change control policies and practices.

2014 The IPCC issues its fifth annual report on climate change, Fifth Assessment Report (AR5).

2015 Executives at six large European oil companies—BG Group, BP, Eni, Royal Dutch Shell, Statoil, and Total—write to Christiana Figueres, Executive Secretary of the United Nations Framework Convention on Climate Change (UNFCCC), asking that action be taken to deal with climate change. Their major suggestion is the imposition of a carbon tax on the use of fossil fuels.

2015 At its annual meeting in Paris, 195 nation-members of the UNFCCC, 196 member-nations adopt the Paris Accord (or Paris Agreement) that lays out a plan for keeping the annual average global temperature increase to less than 1.5°C. Each nation agrees to set a specific target and submit plans to achieve this goal. The United States is one of 195 signatories to the treaty, except that President Donald Trump announces in 2017 that the United States will withdraw from the agreement in 2020.

2015 Obama announces a Clean Power Plan designed to phase out the use of fossil fuels in the country and replace them with alternative sources of energy.

2016 Obama formally enrolls the United States in the Paris Accord.

2017 President Donald Trump (hereinafter, Trump) announces that the United States will withdraw from the Paris Accord on climate change in 2020.

2017 Trump issues an Executive Order rescinding 20 Obama actions on climate change.

2017 Trump announces plans to withdraw the United States from the Paris Accord.

2019 Trump revokes Obama's Clean Power Plan and replaces it with the Affordable Clean Energy rule.

2019 Trump revokes an Obama rule requiring oil and gas companies to reduce methane emissions in their operations.

2019   One of Iceland's most notable glaciers, Okjökull ("Ok") disappears because of extensive melting.

2020   Scheduled date for the withdrawal of the United States from the Paris Accord on climate change.

2022   Scheduled release of the sixth IPCC report on climate change, Sixth IPCC Assessment Report (AR6).

Discussions about climate change often make use of technical and quasi-technical terms, without an understanding of which, that discussion can rapidly lose meaning. The glossary of terms provided here, many found in this book itself, and others taken from outside sources, should facilitate the debate among individuals with varying views of one or more aspects of climate change.

Ablation   The process by which snow and ice are lost from a glacier or ice field.

Aerosol   A form of matter that consists of tiny particles that range in size from about 10^{-3} to 10^{-2} micrometers in size.

Albedo   The fraction of light that strikes a surface and is reflected by it.

Algal bloom   A sudden, rapid growth of algae in lakes and coastal ocean waters caused by a variety of factors including, for example, warmer surface waters or increased nutrient levels.

Anomaly   Some factor or data that deviates from that which is expected.

Anthropogenic   Produced as the result of human action.

Atmospheric general circulation models (AGCMs)   General circulation models that deal only with the atmosphere.

Atmospheric window   That portion of the atmosphere's absorption spectrum between about 8 and 13 micrometers (2×10^{-4}

and 5×10^{-4} inches) through which radiation can normally escape.

Axial precession   The tendency of Earth's axis to wobble in space over a period of about 23,000 years.

Best estimate   The extrapolation from a climate model that appears to be the most likely to occur.

B.P. (or b.p.)   An abbreviation for the phrase "before the present," used to indicate periods in Earth's history.

Business as usual   A situation in which individuals, companies, the government, and other entities continue to operate as they have in the past, with no changes made to deal with some existing or anticipated problem, such as climate change.

BYA (or bya)   The abbreviation for a time period "billions of years ago."

Carbon budget   An accounting of the gain and loss of carbon between any two steps in the carbon cycle.

Carbon cycle   The complex series of reactions by which the element carbon passes though Earth's atmosphere, biosphere, hydrosphere, and lithosphere.

Carbon-dioxide equivalent   The combined climatic effect of greenhouse gases other than carbon dioxide, including methane, nitrogen oxides, and CFCs.

Carbon-dioxide removal (CDR)   Any process by which carbon dioxide produced in a system is captured and buried underground or stored in the oceans.

Carbon footprint   A measure of the contribution made to climate change as a result of an individual or organization's actions.

CFC.   *See* **chlorofluorocarbon**

Chlorofluorocarbon (CFC)   An organic compound containing carbon, chlorine, fluorine, and (usually) hydrogen with a wide variety of commercial and industrial applications.

Climate   The sum total of weather conditions for a particular area over an extended period of time, at least a few decades.

Climate model   Any mathematical and/or physical representation that attempts to simulate the behavior of the climate, allowing scientists to make predictions about future climatic conditions.

Climatic optimum   The historical period with the highest prevailing temperatures from the last ice age. Dates differ for the period, depending on the source, but commonly range from about 4,000 to 8,000 years ago.

Combustion   The chemical process of oxidation. "Burning."

Convection   A method by which heat is transferred from one place to another by means of the movement of a heated package of liquid or gas.

Cryosphere   That portion of Earth's surface that is covered by masses of snow and/or ice.

Dendrochronology   The science of determining the age of some object by studying the growth rings in trees.

Eccentricity   The amount by which the orbit of one object around a second object, such as Earth around the Sun, differs from a perfect circle. A measure of the amount of "flattening" of that orbit.

Emission mitigation analysis   Any study that attempts to predict the effects of reducing the release of one or more greenhouse gases into the atmosphere.

Equilibrium   The condition in which two opposing forces exactly balance each other.

Feedback mechanisms   Any series of changes in which the final step in the series results in an outcome that influences the first step of the series.

Forcing factor   Some factor that, which changed, brings about a change in the climate.

General circulation model (GCM)   Any mathematical and/or physical representation that attempts to replicate one or more existing climate conditions, for use in predicting one or more future climate conditions.

Geoengineering   Large-scale interventions in Earth's natural systems to counteract climate change.

Glacial maximum   The greatest extent to which a glacier has extended during its lifetime. Also, glacial maximum or last glacial maximum, as the most recent time during which glaciers had covered their greatest land area. That number varies for various part of the Earth, but is often given as about 26,500 B.P.

Glaciation   Any process by which a large area of land or water is converted with ice and snow for some significant period of time.

Global warming   The process by which Earth's average annual temperature increases by a significant amount over a relatively extended period of time.

Greenhouse effect   A term used to describe the effect on Earth's temperature that results from the capture of heat by molecules of carbon dioxide, water vapor, and other gases in Earth's atmosphere.

Greenhouse gas   Any gas that does not absorb solar radiation but does absorb long-wavelength radiation reflected from Earth's surface.

Heat balance   A mathematical accounting for all the heat that enters and leaves Earth's atmosphere.

Ice age   Any period in Earth's history when significant portions of Earth's surface was covered by glaciers.

Ice core   A cylindrical section of ice removed from a glacier or ice sheet in order to study climate patterns of the past.

Ice sheet   The name given to any glacier that covers more than 50,000 square kilometers (19,300 square miles) to a significant depth.

Industrial Revolution The period in human history, ranging from about 1750 to 1850, when factory-based manufacturing began to replace home-based businesses in England.

Infrared radiation Electromagnetic radiation in the range between about 0.7 micrometer and 1,000 micrometers (3×10^{-5} and 4×10^{-2} inches).

Insolation The amount of solar radiation that falls on a unit area of horizontal surface in a given period of time.

Interglacial period The period of time between two glacial periods, or ice ages.

Little Ice Age A period in Earth's history that lasted from about 1550 to 1850 in the Northern Hemisphere, when temperatures were somewhat colder than usual and glaciers extended over relatively large areas of North America, Asia, and Europe.

Milankovitch (or Milanković) theory A theory developed by Serbian meteorologist Milutin Milanković in the early 1920s explaining the occurrence of ice ages as the result of Earth's eccentricity, obliquity, and axial precession.

MYA (or mya) The abbreviation for "millions of years ago."

Negative emissions Process in which carbon dioxide produced in some reaction is captured and stored underground or in the oceans.

Nitrous oxide A compound of nitrogen and oxygen with the formula N_2O formed during the combustion of fossil fuels, by the breakdown of chemical fertilizers, and by bacterial action in soil.

100-year flood A flood or storm that has a 1 percent probability of occurring in any given year.

Outgassing The escape of gas from a solid or liquid, as occurs when carbon dioxide is released from the surface of seawater.

Paleoclimatology The study of Earth's climates at distance times in the past.

Palynology   The study of plant spores and pollen in the fossil state, as a means of learning about Earth's climate at distant periods in history.

Phenology   The science that deals with changes in the timing of biological events.

Proxy climate indicators   Any type of data that provides information about climatic conditions during some specific period of time, usually when direct measurements cannot be made.

Radiation balance   The difference between the amount of radiation absorbed by Earth's surface and the amount re-emitted as infrared radiation. Also known as **radiation budget**.

Radiation budget   *See* **radiation balance**.

Reservoir   A place at which a resource is stored. A lake, for example, is a natural reservoir for fresh water.

Residence time   The time during which some substance remains in a particular reservoir, such as the time that carbon dioxide normally remains in the atmosphere.

Sea surface temperature   The temperature of the upper 0.5 meter (1.6 feet) of seawater.

Secular carbon-dioxide trend   The steady increase in concentration of carbon dioxide in the atmosphere.

Self-correcting mechanism   Any process that operates in such a way as to reduce or eliminate some trend in the process itself.

Sequestration   The process by which a substance is removed from the free state and tied up in some other material, such as the conversion of carbon dioxide in the atmosphere to carbohydrates during the process of photosynthesis.

Simulation   Any attempt to imitate or reproduce some physical phenomenon using mathematical and/or physical models.

Sink   The final location in which some material is deposited for an extended period of time. For example, the oceans are a sink for carbon dioxide because the gas dissolves in seawater and tends to remain there for long periods of time.

Solar constant The rate at which solar energy strikes the outermost layer of Earth's atmosphere. The current value of the solar constant is 0.140 watts per square centimeter.

Solar cycle The periodic change in the number of sunspots. The current average interval between successive minima and maxima is about 11.1 years.

Solar radiation management (or **modification; SRM**) Any system by which solar radiation reaching or reflected from Earth's surface is reduced.

Sunspot A region on the Sun's surface that appears to be darker, and therefore cooler, than the surrounding area.

Technological fix An attempt to solve a problem by some technical means, rather than by instituting social, economic, political, or other type of change.

Terrestrial radiation The total amount of radiation emitted by Earth, including its atmosphere, in the temperature range of about –70°C to about 50°C (about –90°F to about 90°F).

Tilt The orientation of Earth's axis in space, compared to the orbital plane, currently about 23.5°.

Tipping point A time period beyond which climate changes that have already occurred are reversible, regardless of any efforts to ameliorate the process of global warming.

Trace gas Any one of the less common gases found in Earth's atmosphere, such as oxides of nitrogen, ozone, and ammonia.

Transpiration The process by which plants give off water from their leaves.

Tree ring The amount by which a tree grows in circumference in a single year.

Ultraviolet radiation Electromagnetic radiation with a wavelength of about 4 to 400 nm (nanometers).

Upwelling   The vertical movement of ocean water that brings cold subsurface water to the surface.

Urbanization   The movement of people from rural areas to regions of more concentrated population, such as cities.

Vapor   The gaseous form of a material that is normally solid or liquid.

Visible light   A form of electromagnetic radiation with a wavelength of about 370 to 730 nm (nanometers).

Weather   The state of the atmosphere as defined primarily by six factors: temperature, barometric pressure, wind velocity and direction, humidity, clouds, and precipitation.

X-rays   A form of electromagnetic radiation with wavelengths of about 1 to 10 nm (nanometers).

Note: Page numbers followed by *t* indicate tables and *f* indicate figures. Page numbers in *italics* indicate photos.